Indoor air quality
design guidebook.

$62.00	DATE		

INDOOR AIR QUALITY DESIGN GUIDEBOOK

INDOOR AIR QUALITY DESIGN GUIDEBOOK

Edited by

Milton Meckler, P.E.

Published by
THE FAIRMONT PRESS, INC.
700 Indian Trail
Lilburn, GA 30247

Library of Congress Cataloging-in-Publication Data

Milton Meckler, P.E.
Indoor air quality design guidebook / edited by Milton Meckler.
p. cm.
Includes bibliographical references.
ISBN 0-88173-088-2
1. Indoor air pollution. 2. Air quality management.
I. Meckler, Milton.

TD883.1.I48 1991 ~~1990~~ 697-dc20 89-45175
CIP

Indoor Air Quality Design Guidebook.

Published by The Fairmont Press, Inc.
700 Indian Trail
Lilburn, GA 30247

Printed in the United States of America

10 9 8 7 6 5 4 3 2 1

ISBN 0-88173-088-2 FP

ISBN 0-13-457284-X PH

Distributed by Prentice-Hall, Inc.
A division of Simon & Schuster
Englewood Cliffs, NJ 07632

Prentice-Hall International (UK) Limited, London
Prentice-Hall of Australia Pty. Limited, Sydney
Prentice-Hall Canada Inc., Toronto
Prentice-Hall Hispanoamericana, S.A., Mexico
Prentice-Hall of India Private Limited, New Delhi
Prentice-Hall of Japan, Inc., Tokyo
Simon & Schuster Asia Pte. Ltd., Singapore
Editora Prentice-Hall do Brasil, Ltda., Rio de Janeiro

CONTENTS

PART 1
SOURCES OF INDOOR AIR POLLUTION AND HEALTH EFFECTS

PART 2 ENGINEERING SOLUTIONS TO INDOOR AIR QUALITY PROBLEMS

PART 3
PREVENTIVE INDOOR AIR QUALITY MEASURES

CONTRIBUTORS

REN ANDERSON, Technology Leader, Solar Energy Research Institute, Building Research Branch, Chapter 9.

PETER W. H. BINNIE, Vice-President, ACVA Atlantic, Inc., Chapter 13.

HOWARD GOODFELLOW, President, Goodfellow Consultants Inc., Chapter 10.

LARRY C. HOLCOMB, PH.D, President, Holcomb Environmental Services, Chapters 1, 4 and 5.

CARL N. LAWSON, Vice-President, Liebtag, Robinson & Wingfield, Inc., Chapter 14.

MILTON MECKLER, P.E., President, The Meckler Group, Chapters 3, 7, 11 and 15.

DEMETRIUS J. MOSCHANDREAS, PH.D., Sr. Science Advisor, IIT Research Institute, Chapter 12.

SERESH RELWANI, Research Engineer, IIT Research Institute, Chapter 12.

ELIA M. STERLING, Director of Building Research, Theodor D. Sterling and Associates, Ltd., Chapters 1, 2, 3, 4, 5, 6 and 7.

FRANK VACULIK, Sr. Operation & Maintenance Engineer, Public Works Canada, Chapter 8.

FOREWORD

The primary purpose of a building is to provide a healthy and comfortable indoor environment for its occupants. Recent studies show that indoor air in homes, schools, offices and other non-residential buildings contains solvents, chemicals, gases, smoke, bacteria and other pollutants, sometimes in high enough concentrations to pose a serious threat to our health.

Growing concern about indoor air quality (IAQ) problems may be largely due to: the introduction of pollutants into the buildings by newly developed synthetic materials; some energy conservation measures resulting in diminished outside fresh air; tightly sealed buildings preventing pollutants from escaping; lower quality design, construction, operation and maintenance; and the tendency toward increased litigation necessitating preventive measures.

IAQ has become a critical consideration in designing heating, ventilating and air-conditioning (HVAC) systems. It is the responsibility of an HVAC designer to select a system that would ensure effective operation and obtain the necessary commitment toward maintaining a healthy and comfortable indoor environment. Although preventive and corrective measures may, in some cases, increase the construction and occupancy costs, inadequate IAQ measures can be more costly in terms of loss of productive time and in the removal and/or replacement of materials to correct the problem.

The intent in preparing this book was to bring together experts in their respective fields in a collaborative effort to deal with the health effects in humans, and engineering solutions to IAQ problems. The Indoor Air Quality Design Guidebook is a consolidated book for use by architects, engineers, contractors, building owners, buildings manager, and building operators who need to provide a healthy and productive environment for all building occupants.

The material is presented in three parts. Part 1 (Chapters 1-6) reviews the potential sources of indoor air pollution and health effects based on most recently researched and published data. Part 2 (Chapters 7-12) covers engineering solutions to IAQ problems in the areas of pollutant control, evaluation of methods for measuring major indoor air pollutants, techniques for modeling ventilation efficiency, IAQ simulation by computer modeling, zoning for IAQ, and the role of solid desiccants in IAQ control. Materials presented in these chapters reflect state-of-the-art solutions to IAQ problems which are intended to assist in finding viable alternative solutions to complex environmental concerns in accordance with the currently approved design standards (e.g., ASHRAE Standard 62-1989). Finally, Part 3 (Chapters 13-15) deals with the often overlooked preventive aspects of IAQ, namely: system operation and maintenance, commissioning, and the investigation of buildings which have reported sick-building syndrome (SBS) or building-related illness (BRI).

Sincere thanks are due to our contributors who shared their valuable time, immense knowledge, and patience to make this book possible. Special thanks are also due to Refik A. Sar, Director of Communications at The Meckler Group, who provided invaluable assistance to me in the preparation of this material.

Milton Meckler, P.E.
President
The Meckler Group

PART 1

SOURCES OF INDOOR AIR POLLUTION AND HEALTH EFFECTS

1. FORMALDEHYDE

Elia M. Sterling
Director of Building Research, Theodor D. Sterling and Associates, Ltd.
Vancouver, B.C., Canada

and

Larry C. Holcomb, Ph.D.
President, Holcomb Environmental Services
Olivet, Michigan

1.1 Sources and Concentrations

Formaldehyde belongs to the chemical group of aldehydes. It has the chemical formula HCHO, and is highly reactive and soluble in water. Formaldehyde may be found in three physical states, as a pure gas, an aqueous solution, and a solid polymer (1). Since formaldehyde is highly soluble in water, it can be irritable to all body surfaces normally containing moisture, such as the eyes and upper respiratory tract.

Formaldehyde is a major component of urea-formaldehyde-foam insulation (UFFI), particle board, some paper products, fertilizers, chemicals, glass, and packaging material. Industrial emissions are a major outdoor source of formaldehyde. The incomplete combustion of hydrocarbon-based fuels used in transportation is estimated to produce approximately 666 million pounds of formaldehyde annually in the U.S. (2).

Some office environments may be exposed to measurable formaldehyde concentrations because of off-gassing from particle boards used in furniture and wall coverings. Measurements in offices have found average concentrations of formaldehyde between 0.04 ppm and 0.06 ppm (3). Indoor formaldehyde concentrations have been found to be substantially higher than in other occupational environments. For example, autopsy rooms have yielded mean values of 1.58 ppm (4). Concentrations of formaldehyde as high as 1.56 ppm have been observed in shopping centers (5). Industrial workers are usually exposed to less than 1 ppm of formaldehyde, though plywood and particle board workers have been exposed to mean concentrations of 1 ppm to 2.5 ppm (5).

The major indoor source of formaldehyde has been from foam insulation containing formaldehyde itself and particle board. Manufactured housing, particularly mobile homes, have often been found to have substantial concentrations of formaldehyde in the air due primarily to the extensive use of particle board (chipboard) and plywood in construction, tight sealing, and the increased use of recirculated air. In addition, the simple presence and normal activity of humans have been found to triple the amount of aldehydes indoors (6). Concentrations of formaldehyde measured in mobile homes have been between 0.01 ppm and 3 ppm (7). Increase in mobile home use has been substantial in the sunbelt regions such as Texas, where high temperatures and humidity may further enhance off-gassing of formaldehyde.

One of the potential sources of formaldehyde contamination is UFFI. Houses with UFFI have yielded formaldehyde concentrations ranging from 0.01 ppm to 4 ppm (8) compared to conventional homes which may have formaldehyde concentrations between 0.02 ppm and 0.4 ppm (9, 10). If UFFI is properly installed, the high concentrations of formaldehyde generation which occur as a result of the polymerization process cease fairly rapid. On the other hand, improper installation can cause high concentrations of formaldehyde generation for long periods of time. Other sources of formaldehyde in the residential and commercial environments include floor coverings, carpet backings, adhesives, and particle boards. Testing of carpeting and particle board samples removed from the paired experimental mobile homes has shown that, in general, temperature has a greater effect on formaldehyde emission rates than relative humidity.

1.2 Health Effects

The wide range of concentrations at which symptoms occur suggest a large variation in individual sensitivity to formaldehyde (11). According to acute irritation studies, most people feel a burning sensation in their eyes, noses, and throats at exposures between 0.1 ppm and 3 ppm. In chronic exposures of 2 ppm or higher, formaldehyde can begin to destroy the lining of the nose, diminishing the ability of the respiratory system to clean out airborne particles and microbes and lead to other respiratory illnesses. Table 1-1 summarizes the level at which various effects are reported to occur.

Table 1-1. Acute Human Health Effects of Formaldehyde at Various Concentrations (11).

Reported Effect	Concentration (ppm)
None reported	0.0 - 0.05
Odor threshold	0.05 - 1.0
Eye irritation	0.01 - 2.0
Upper airway irritation	0.1 - 25
Pulmonary effects	5 - 30
Pulmonary edema, inflammation	50 - 100
Death	> 100

According to a recent study which measured the acute effects of formaldehyde in nonsmokers (12), the eye and nose/throat irritations increased as formaldehyde concentrations were increased from 0.5 ppm to 3 ppm. Nasal resistance increased significantly at a concentration of 3 ppm but no significant changes in pulmonary function or reactivity were observed. Symptoms of wheezing, headaches, fatigue, and reduced lung function have been also attributed to both residential and occupational exposures to formaldehyde.

Formaldehyde has been shown to produce nasal cancer in laboratory rats (11). Data on carcinogenesis in humans, though, are inconclusive. Twenty-eight epidemiologic studies of cancer in residential and occupational exposures to formaldehyde have been reviewed by the Environmental Protection Agency (EPA). Nine of these studies have shown a significant increase in respiratory tract cancer related to reported exposure with relative risks ranging from 1.6 to 2.5. These studies, though, have been complicated by possible exposure to a wide variety of other chemicals. Although the International Agency for Research on Cancer (IARC) stated that there is a limited evidence of carcinogenicity, the EPA felt that the evidence was sufficient enough to declare formaldehyde a "probable carcinogen." An additional evidence for the carcinogenic potential of formaldehyde is its reported ability to produce breaks in DNA chains, mutations, and chromosonal alterations (13,14).

Studies on the epidemiology of formaldehyde have been mainly restricted to cohorts of occupationally exposed individuals in the plastics industry and of undertakers, embalmers, anatomists, pathologists, and medical technologists involved in anatomical or cytological studies. Other professions include the materials production, manufacturing, and finishing industries where formaldehyde resins are used. Occupational exposure levels typically range from 0.1 ppm to 5 ppm (11).

Statistically significant proportional mortality ratios were observed for buccal and pharyngal cancer among workers in a large chemical plant (15). This is of particular interest because of the close similarity between animal studies which found an increased incidence of nasal cancer. There are no other epidemiological studies which found an increased incidence of cancer of the upper breathing passages, but there are reports of nasal cancer among people formally exposed to formaldehyde in industrial shops (16, 17). However, nasal cancer is also related to other shop exposures especially to soldering, welding, and flame cutting. Another recent study found elevated oral and hypopharynx cancers in relation to occupational formaldehyde exposure (18). Cancer of the nasal pharynx was also found to be related to residential formaldehyde exposure (19).

Elevated risk for cancer of the respiratory region has been reported for garment workers who were exposed to formaldehyde (20) and for industrial workers (21). Reanalysis of a British study of chemical workers (22) and controlling for the healthy-worker-effect by computing proportional-mortality- ratios also found elevated lung cancer rates (23). A joint study by the U.S. National Cancer Institute and Formaldehyde Institute found elevated lung cancer risks which were however not statistically significant (24). However, a reanalysis of these data that simultaneously control for the healthy-worker-effect found elevated lung cancer risks for the same population (25). It may be concluded that exposure to certain concentrations of formaldehyde probably increases the risk for lung cancer and cancer of the upper respiratory region.

Following formaldehyde exposure, cancers of the following areas have been also reported: the digestive system (26), stomach and pancreas (27), colon (15, 27), bladder (27), kidney (26, 28), skin (26), and cancer of the lymphatic and haemopoetic systems (26, 29, 30). Finally, the fact that formaldehyde combines with hydrogen chloride to form bischloromethylether, which is known to cause small cell undifferentiated lung cancer in humans (31), further suggests the carcinogenic potential of formaldehyde.

In light of the evidence, it would be reasonable to conclude that there may be an increased risk of cancer from long-term exposure to formaldehyde. Obviously, the amount of risk would depend on the amount of formaldehyde present in the indoor environment. Short-term formaldehyde concentrations in some residential buildings have been as high as in those workplaces where a possible association between formaldehyde exposure and cancer has been found. Indoor formaldehyde concentrations because of off-gassing from new materials tend to decrease with time and probably disappear after a year or two depending on humidity, temperature, and other factors (32, 33). This suggests that indoor formaldehyde exposures may be too short to constitute a significant cancer risk except for a small percentage of the population.

References

1. Sterling, T.D. and Arundel, A. "Possible Carcinogenic Component of Indoor Air: Combustion Byproducts, Formaldehyde, Mineral Fibers, Radiation, and Tobacco Smoke." *J. of Environ. Sci. and Health*, C2(2), pp. 185-230, 1984.

2. Gibson, J.E. (Ed.). Formaldehyde Toxicity, Hemisphere, NY, 1983.

3. Blade, L.M. "Formaldehyde Toxicity Epidemiology Mechanisms." Edited by Clary, J.J.; Gipson, J.E.; and Waritz, R.S., p. 1, NY, 1983.

4. Makar, A.B.; McMartin, K.E.; Palese, M.; and Tephly, T.R. Biochemical Medicine, 13, p. 117, 1975.

5. National Institute for Occupational Safety and Health. "Industry Selection for Determination of Extent of Exposure." NIOSH Center for Disease Control, Cincinnati, OH, 1979.

6. Hollowell, C.D.; Berk, J.V.; and Lin, C.I. "Indoor Air Quality in Energy-Efficient Buildings." Lawrence Berkeley Laboratory, Energy and Environment Division, University of California, Berkeley. Report No. LBL 8892, EEB Vent 79-2, 1979.

7. National Academy of Sciences. "Indoor Pollutants." National Academy Press, Washington, DC, 1981.

8. Gupta, K.C.; Ulsamer, A.G.; and Preuss, P.W. Environ. Int., 8, p. 349, 1982.

9. Hawthorne, A.R.; Gammage, R.B.; Dudney, C.S.; Cormack, D.R.; Morris, S.A.; and Wesley, R.R. Proceedings of the Conference on Measurement and Monitoring on Non-Criteria (Toxic) Contaminants in Air, p. 514, Air Pollution Control Association, 1983.

10. Nero, A.V. and Grimsrud, D.T. "Dependence of Indoor Pollutant Concentration on Sources, Ventilation Rates and Other Removal Factors." Lawrence Berkeley Laboratory Report. LBL-16525, Berkeley, CA, 1983.

11. Samet, J.M.; Marbury, M.C.; and Spengler, J.D. "Health Effects and Sources of Indoor Air Pollution." Part II, *Am. Rev. Respir. Dis.*, 137, pp. 221-242, 1988.

12. Kulle, T.J.; Sauder, L.R.; Hebel, J.R.; Green, D.J.; and Chatham, M.D. "Formaldehyde Dose-Response in Healthy Nonsmokers." *JAPCA*, 37, pp. 919-924, 1987.

13. Woutersen, R.A.; Appelman, L.M.; Feron, V.J.; and Van der Heijden, C.A. Toxicology, in press.

14. Grafstrom, R.C.; Fornace, A.J.; Autrup, H.; Lechner, J.F.; and Harris, C.C. Science, 220, p. 216, 1983.

15. Liebling, T.; Rosenman, K.D.; Pastidies, H.; Griffith, R.G.; and Lemeshow, S. Am. J. Ind. Med., 5, p. 423, 1984.

16. Infante, P. and Kand, H. "Nasal Sinus, Pharyngal and Buccal Cavity Cancer and Formaldehyde Exposed Workers." OSHA internal document, Washington, DC, 1982.

17. Halperin, W.E.; Goodman, M; and Stayner, L. JAMA, 249, p. 510, 1983.

18. Vaughan, T.L.; Strader, C.; Davis, S.; et al. "Formaldehyde

and Cancers of the Pharynx, Sinus, and Nasal Cavity: I. Occupational Exposures.'' *Int. J Cancer*, 38, pp. 677-684, 1986.

19. Vaughan, T.L.; Strader, C.; Davis, S.; et al. ''Formaldehyde and Cancers of the Pharynx, Sinus, and Nasal Cavity: II. Residential Exposures.'' *Int. J Cancer*, 38, pp. 685-688, 1986.

20. Stayner, L.; Smith, A.B.; Reeve, G.; et al. ''Proportionate Mortality Study of Workers in the Garment Industry Exposed to Formaldehyde.'' *Am. J Ind. Med*, 7, pp. 229-240, 1986.

21. Wong, O. ''An Epidemiologic Mortality Study of a Cohort of Chemical Workers Potentially Exposed to Formaldehyde with a Discussion on SMR and PMR.'' Edited by Gibson, J.E. *Formaldehyde Toxicity, Chemical Industry Institute of Toxicology Series*, Washington, DC, pp. 256-272, 1983.

22. Acheson, E.D.; Barnes, H.R.; Gardner, M.R.; et al. ''Formaldehyde in the British Chemical Industry.'' *LANCET*, pp. 611-616, 1984.

23. Sterling, T. and Arundel, A. ''Formaldehyde and Lung Cancer.'' *LANCET*, pp. 1366-1367, 1985.

24. Blair, A.; Stewart, P.; O'Berg, M.; et al. ''Mortality Among Industrial Workers Exposed for Formaldehyde.'' *JNCI*, 76, pp. 1071-1084, 1986.

25. Sterling, T.D. and Weinkam, J.J. ''Reanalysis of Lung Cancer Mortality in a National Cancer Institute Study on Mortality Among Industrial Workers Exposed to Formaldehyde.'' *J of Occ. Med.*, 30, pp. 895-901, 1988.

26. Walrath, J. and Fraumeni, J.F. Int. J Cancer, 31, p. 407, 1983.

27. Marsh, G.M. Formaldehyde Toxicity. Edited by Gibson, J.E. Hemisphere, NY, 1983.

28. Tabershaw Associates, Inc. ''Historical Prospective Mortality Study of Past and Present Employees of the Celanese Chemical and Plastics Plant Located in Bishop, Texas.'' 1982.

29. Levine, R.J.; Andjelkovich, D.A.; and Shaw, L.K. *Formaldehyde Toxicology, Epidemiology and Mechanisms*. Edited by Clary, J.J.; Gibson, J.E.; and Waritz, R.S., p. 127, NY, 1983.

30. Harrington, J.M. and Shannon, H.S. Brit. Med. J., 2, p. 329, 1974.

31. United States Department of Health, Education and Welfare. Occupational Diseases: A Guide to Their Recognition, DHEW (NIOSH) publication No. 77-81 (2nd printing), Sept., 1978.

32. Sterling, D.A.; Stock, T.H., and Monteith, D.K. Indoor Air, Vol. 3. Edited by Berglund, B.; Lindvall, T.; and Sundell, J., p. 139, Swedish Council of Building Research, Liber Tryck AB, 1984.

33. Heinberg, S.; Collan, Y.; and Degerth, R. Scand. J. Work Environ. Health, 9, p. 208, 1983.

2. RADON

Elia M. Sterling
Director of Building Research, Theodor D. Sterling
and Associates, Ltd.
Vancouver, B.C., Canada

Page

2.1 Sources and Concentrations

Because individuals spend the majority of their lives indoors, they have always been exposed to radiation from natural sources. The main source of natural radiation exposure is the distribution of the radioactive elements such as uranium and thorium in the bedrock and soils. All types of rocks and soil contain these radioactive elements in very low concentrations. Although low-level exposure to radiation will not produce any immediate health hazards, high-level or prolonged exposure may result in serious health problems such as lung cancer. There are no early visible warning signs of cancer that may develop 10 to 20 years following exposure.

One of the particular sources of radiation is radon gas. Radon is an odorless, colorless, radioactive gas which is the decay product of radium-226 and, radium-226 is the decay product of uranium-238. Radioactive elements such as uranium and radium are further distributed in rock products, and by chemical precipitation of groundwater. Unlike other elements from uranium to lead in the radioactive decay series of chemical chart, radon is a chemically inert (inactive) gas. When radium decays to radon, it drifts and moves through the rocks and soil totally unattracted to other elements. Some radon enters the air before even decaying into its four short-lived daughter products. Two of the four radon daughter products with a short half-life of approximately 30 minutes emit alpha particles (positively charged helium nuclei) and the other two emit beta particles (electrons) when they decay. Furthermore, these particles may separate electrons from their atoms and create electrically charged particles called ions.

The concentration of radon in the air is measured in picoCuries per liter of air (pCi/L) (1). One pCi is equal to 0.037 nuclear disintegration per second. Human exposure to radon is measured in Working Levels (WL) or Working Level Months (WLM). One WL is equivalent to 100 pCi/L of radon at equilibrium with its daughters. One WLM equals to an exposure of 680 WL hours (2). The equilibrium factor, or the proportion of radon daughters predicted on the basis of decay rates, is generally assumed to be 0.5 for residences but is known to vary between 0.3 and 0.8 (3). Radon daughters do not occur at the predicted equilibrium with radon due to the loss of daughters attached to surfaces or to settling aerosols (4).

Presently in the U.S., the occupational standard for exposure to radon daughters is 4 WLM (1). The equivalent of exposure to the occupational standard may occur during long-term exposure to comparatively low concentrations of radon. Exposure to an average of 0.12 WL for 66% of the year will also result in exposure to the occupational standard of 4 WLM per year. Presently, the Environmental Protection Agency (EPA) recommends remedial action to reduce indoor residential exposures if concentrations exceed 0.015 WL (5).

The overwhelming source of indoor radon is the soil. It is transported from soil through open sumps, crawl spaces, hollow concrete block walls, and cracked concrete slabs. Recent evidence further suggests that the dominant transport process by which radon enters the building is the flow due to pressure differential rather than molecular diffusion. For example, two air masses of different temperatures separated by a vertical wall create a pressure differential that is a function of height. During the hot season, air due to gravity gives rise to a net pressure differential between the floor and lower parts of the house walls, thus creating a flow that passes through the soil adjacent to the building. Wind is the other factor which generates a pressure differential due to an exchange of air volume between the house and the soil.

Radon also enters buildings from the outdoor air and crustal building materials such as concrete, brick, stone, or drywall made of phosphogypsum. Indoor radon concentrations depend on both the rate of entry from source materials and factors such as ventilation and aerosol settling rates. Human exposure to radon is also strongly influenced by the time spent indoors and the location of the indoor activity. Indoor radon concentrations are consistently higher in basements and inversely proportional to the height. Therefore, individuals in basement apartments may be exposed to comparatively higher radon concentrations than those at the first or second floors. In multistory buildings, the most important source of radon is the building materials (6). Radon exhalation rates are the highest for concrete and brick and lowest for wood, other organic building materials, and steel. In the U.S., the average exhalation rates for concrete are between 0.4 and 1.2 pCi/Kg/hr, which add approximately 0.0002 WL to other sources of exposure.

2.2 Health Effects

The primary health hazard from exposure to radon and its daughters is an increased risk of lung cancer. The most significant health risk comes from two of the four radon daughters that may be inhaled into the lungs directly or by airborne particles. Two of the radon daughters decay by the emission of alpha particles. The main health hazard is due to the alpha particles that have a shorter range. These particles deposit more energy in a smaller section of a human tissue causing increased risk of cancer. Radon daughters, when inhaled into the lungs, may lodge in the bronchial tubes. Radioactive particles may also cross from lungs into the bloodstream or be swallowed with expectorated mucous and cause internal exposure leading to cancer of other organs.

The relationship between the radon exposure and cancers of the lung and stomach has been conclusively shown for uranium miners in the U.S. (7, 8, 9). Exposure data for underground miners have been relatively good and permitted the development of estimates of excess cancer deaths for a given exposure. Estimates range from 6 to 19 deaths per one-million-person years/WLM (10). The variation in excess risk estimates is most likely due to variations in the accuracy of exposure data and the length of follow-up for mining cohorts. In addition to miners, elevated lung cancer risks have been found for occupants of houses built of stone (11).

Table 2-1 shows the cumulative exposure to radon daughters for uranium miners and the general public. The general public is assumed to spend 80% of its time indoors and exposed to average concentrations of radon daughters. Assuming an incidence rate of 5 per million persons/WLM and a cumulative exposure of 10 WLM, then the total U.S. population (over 200 million) can be expected to have about 10,000 cases of lung cancer per year (12) due to radon exposures.

Table 2-1. Exposure to Radon Daughters.

Population or Location	Exposure (WLM)
Uranium miners	100-1000
Outdoors	< 1
Indoors	10

Source: National Research Council, Indoor Pollutants (Washington, DC, National Academy Press, 1981)

The risk of getting lung cancer increases as the radiation dose increases. To assess the health effects at low-level exposures, it is common to assume that the occurrance of lung cancer is directly proportional to the cumulative dose from the decay products of radon.

2.3 Factors Affecting Concentration of Radon and Control Techniques in Buildings

Indoor radon concentrations are directly related to the ventilation or fresh air exchange rates measured in air changes per hour (ach). However, employing ventilation in houses has not always led to a significant or meaningful correlation due to wide variations in the rate of entry of radon into the indoor environment. On the other hand, for a given rate of radon entry, indoor radon concentrations are higher in energy-conserving houses with low ventilation rates below 0.5 ach than for conventional houses between 0.5 ach and 1.5 ach. According to a study conducted in New York (5) involving 21 energy-efficient and 14 conventional houses, indoor radon concentrations in energy-efficient homes have been found to be approximately 2.7 pCi/L yearly as compared to 0.89 pCi/L for conventional homes.

Another important factor that may have a significant effect on indoor radon concentrations is the radon gas dissolved in water. Radon concentrations in the U.S. water supplies were below 2000 pCi/L for 74% of the samples tested but exceeded 10,000 pCi/L for 5% of the samples (2). Each increase of 10,000 pCi/L in water is estimated to increase indoor radon concentrations by 0.8 $\pm$0.2 pCi/L of air (13). Radon dissolved in water is released into the air largely through activities that create water turbulence such as showering and washing of dishes or clothes.

Several controlling techniques can be utilized if indoor radon concentrations of more than a few pCi/L are detected indoors. These techniques to control concentrations of radon and its daughters include measures that: (a) decrease radon sources, (b) reduce transport of radon from its sources to the indoors, (c) remove radon and its daughters from indoor air, and (d) exchange inside air with outside air.

2.4 Technology in New Construction and Cost Effective Methods in Existing Structures for Radon Reduction

As the improvement in construction techniques and building materials becomes more widespread, radon gas emanating into and becoming entrapped in residential buildings appears as an increasingly threatening health hazard. Older buildings, especially wood-frame residences, allow indoor air to exfiltrate and outdoor air to infiltrate through cracks in floors and ceilings and around window and door frames, and through other openings in the building structure (14). Contemporary energy-efficient residential technology has allowed the construction of tightly sealed buildings, eliminating material infiltration and exfiltration. Therefore, it is important that residences be designed and constructed to prevent the entry of radon.

The ventilated crawl spaces, combined with vapor barriers in bottom floors, provide an effective means to reduce indoor radon concentrations. One previous study (15) found that the use of ventilated crawl spaces and air/vapor separation of crawl spaces and occupied area reduced indoor radon concentrations by approximately 60%. The same study was expanded to investigate the effect of mechanically ventilated crawl spaces. Compared to a non-ventilated crawl space, mechanical ventilation of a crawl space led to an approximate reduction of 70% in indoor radon concentrations.

When new buildings are constructed, careful selection of building materials can be very effective in reducing the possible sources of radon. The selection of a construction site is often very difficult. Nevertheless, special measures must be taken to reduce the transport of radon from soil to indoors. This is a less costly approach than soil removal and also less permanent.

References

1. Sterling, T.D. and Arundel, A. "Possible Carcinogenic Components of Indoor Air: Combustion Byproducts, Formaldehyde, Mineral Fibers, Radiation, and Tobacco Smoke." *J. of Environ. Sci. and Health*, C2(2), pp. 185-230, 1984.

2. Nero, A.V. Health Phys., 45, p. 303, 1983.

3. Stranden, E.; Berteig, L.; and Ugletveit, F. Health Phys., 36, p. 413, 1979.

4. Porstendorfer, J.; Wicke, A.; and Schraub, A. Health Phys., 34, p. 465, 1978.

5. Fleischer, R.L. and Turner, L.G. Health Phys., 46, p. 999, 1984.

6. Steinhausler, F.; Hofmann, W.; Pohl, E.; and Pohl-Ruling, J. Health Phys., 45, p. 331, 1983.

7. Axelson, O. "Epidemiology of Occupational Cancer, Mining and Ore Processing." *ILO, Occupational Safety & Health Series*, No. 46, pp. 135-149, 1982.

8. Archer, V.E. and Wagoner, J.F. Health Phys., 25, p. 351, 1973.

9. Samet, M.; Kutvirt, D.M.; Waxweiler, R.J.; and Key, C.R. N. Engl. J. Med., 310, p. 1481, 1984.

10. Radford, E.P. and St. Clair Renard, K.G. N. Engl. J. Med., 310, p. 1485, 1984.

11. Axelson, O. and Edling, C. "Health Hazards from Radon Daughters in Dwellings in Sweden." *Rom WN*. Edited by Archer, V.E. Health Implications on New Energy Technologies. Ann Arbor Science Publishers Inc., Ann Arbor, MI, pp. 79-87, 1980.

12. Turiel, I. Indoor Air Quality and Human Health, 1985.

13. Hess, C.T.; Weiffenbach, C.V.; and Norton, S.A. Health Phys., 45, p. 339, 1983.

14. Sterling, T.D.; McIntyre, E.D.; and Sterling, E.M. "Reducing Radon Levels in Tightly Sealed Residences Through Crawl Space Mechanical Ventilation." *Practical Control of Indoor Air Problem*, IAQ '87, 1987.

15. Sterling, T.D.; Arundel, A.; McIntyre, D.; and Sterling, E. "Effectiveness of Air Vapor Barriers Combined with Ventilated Crawl Spaces in Decreasing Residential Exposure to Radon Daughters: Preliminary Report." Proceedings, Indoor Radon, Air Pollution Association, Philadelphia, PA, pp. 142-147, 1986.

3. PARTICULATES

Milton Meckler, P.E.
President, The Meckler Group
Encino, California

and

Elia M. Sterling
Director of Building Research, Theodor D. Sterling
and Associates, Ltd.
Vancouver, B.C., Canada

3.1 Introduction

Unlike large particles, small particulates may penetrate into the lungs causing mild to serious health problems. The size and density of a particulate with carcinogenic characteristics are the determining factor of how deep it can penetrate the respiratory system, and to what extent its damage will be. In this chapter, our discussion of particulates will be confined to asbestos, fiberglass, rock wool, viruses, bacteria, and allergenic agents. Particulates due to environmental tobacco smoke (ETS) are discussed in Chapter 5, Part 1.

3.2 Asbestos

Asbestos belongs to a group called mineral fibers. It has been used for insulation and fireproofing. Asbestos is a major component of vehicle brakes which produce a constant source of ambient asbestos fibers. Other outdoor sources include asbestos mines and factories. Indoor asbestos contamination may occur by the breakdown of friable insulation or during renovation and construction. It may also occur due to the use of asbestos vinyl floor tiles and during drywalling and spackling taken place prior to 1977. Electrostatically charged asbestos fibers provide attachment sites for other airborne carcinogens such as radioactive and hydrocarbon particles. Asbestos was rarely used in homes, other than for the purpose of sealing fireplaces or heating duct joints. Therefore, exposure to asbestos in homes is usually limited to outdoor sources such as soils or asbestos fibers brought in by the individuals occupationally exposed to asbestos. According to a survey conducted in Globe, Arizona, homes built on asbestos mine tailings contained only 0.004 fibers/cc during undisturbed conditions, but these levels rose to 0.35 fibers/cc during vacuuming (1).

Asbestos contamination in buildings is now recognized as a significant environmental problem. According to the Environmental Protection Agency (EPA), asbestos-containing materials can be found in nearly one million public and commercial buildings in the U.S. Asbestos was widely used in public buildings in the U.S. between 1940 and 1973 for either insulation or decoration. It was sprayed onto the walls or ceilings (2). Possible concern of health hazards to asbestos exposure of young children has led to several surveys of asbestos contamination in schools. Thirty-two percent of schools in New York City and 11.5% in Massachusetts have been estimated to contain asbestos (3). In 27 samples collected from ten schools with visible damage to asbestos surfaces, the concentrations of asbestos were found to range from 9 to 1950 $\mu g/m^3$ with an average of 217 $\mu g/m^3$. The concentrations of asbestos between 100 $\mu g/m^3$ and 1000 $\mu g/m^3$ have been determined to cause pleural cancer (3). In addition, asbestos concentrations of above 0.1 fiber/cc were found during normal activities in dormitories, schools, and offices containing friable asbestos materials (4).

In buildings, materials containing asbestos are mainly found in three categories (5): (a) surface materials and thermal or acoustic insulation on ceilings and walls; (b) insulation around hot or cold pipes, ducts, boilers and tanks; and (c) finishing materials such as ceiling and floor tiles and wall boards. Buildings in categories (a) and (b) are of greatest concern especially if the product is in a form that can be crumbled, pulverized, or reduced to powder by hand pressure. However, the presence of asbestos in a building does not always endanger the health of its occupants. If materials containing asbestos are maintained in good repair and not disturbed, danger to health is unlikely. Asbestos becomes a health hazard only when fibers are released into the air and become friable. Release of fibers in quantity occurs mainly during maintenance, repair, renovation or other construction activities, but fibers may also be released at constant low rates in a process of slow decay.

To reduce the threat of asbestos contamination in buildings, a program containing asbestos inspection, maintenance, and control should be implemented (5). This program should incorporate:

- A survey to identify asbestos containing materials in spaces occupied by tenants;

- Inspection of all areas suspected of containing asbestos to determine the location and condition of asbestos and, if there are sources of friable asbestos;

- Determination of airborne asbestos levels where the urgency of removal is indicated; and

- Preparation of specifications to remove asbestos from those buildings where removal is required. Upon removal of asbestos, control should be part of the normal operations and maintenance.

Conclusive evidence is now available for the carcinogenecity of asbestos. Exposure to asbestos has been definitely linked to stomach and lung cancers, mesothelioma (cancer of the pleura or peritoneum), and fibrotic lung diseases (6). Data on low exposure for occupationally and residentially exposed individuals are generally insufficient to intelligently predict the number of cancer cases. However, studies on individuals who live with asbestos workers or near asbestos factories indicate that mesothelioma may be caused by very low exposure. Several studies conducted on asbestos miners (7), insulation workers (8) and factory workers (9) have indicated an increase in lung, gastrointestinal, pancreatic, and pleural cancers. Another study on radiographic lung abnormalities among the family members of asbestos workers found a significant increase in pleural abnormalities among individuals who did not become household residents until after the end of the asbestos worker's employment (10). This suggests that asbestos contamination in buildings may occur for an extended period of time, and that very low exposures may lead to cancer.

Mesothelioma as a result of neighborhood exposure to outside asbestos sources from local factories or mines has been noted in case reports (11) and epidemiological studies (12). A review of 812 reported cases of mesothelioma in ten countries has linked 18.8% of the cases to nonoccupational exposure as a result of either the location of residence or occupation of a family member (13).

3.3 Fiberglass and Rock Wool

Fiberglass and rock wool have been used mainly for insulation behind walls or other protective coverings in contrast to asbestos frequently sprayed directly onto exposed indoor surfaces (14). Therefore, an average exposure to fiberglass or rock wool is likely to be much less than the average exposure to asbestos during normal activities.

The carcinogenecity of fiberglass and rock wool is currently under investigation because of their physical similarities to asbestos. Some studies indicate lung tumors in animals as a result of implanting fiberglass. Fiberglass can cause severe skin irritations in people if long sleeves and gloves are not worn. It can also be irritating to the eyes and, if inhaled, to the upper respiratory tract. Epidemiological studies have not yet demonstrated substantial health hazards related to fiberglass. Nevertheless, exposure to small fiberglass should be minimized to avoid irritation.

3.4 Viruses, Bacteria and Allergenic Agents

Most airborne viruses and bacteria are generated indoors by occupants. They are transmitted by three routes: airborne, fomite, or direct transmission (15). The airborne transmission occurs when small size particulates with 1 to 5 microns in diameter (infective aerosols) are inhaled and deposited on the surfaces of the tracheal-bronchial area as shown in Figure 3-1. Smaller particulates are deposited in the lungs. As a result of these physical conditions, substances absorbed on the surface of particles can exert potentially harmful effects at the point of deposition of particles on which they are transported. Viruses and bacteria in the lungs, throat, and saliva can be aerosolized by coughing and, therefore transmitted airborne. The airborne transmission of infectious diseases indoors depends on: (a) the number of infected people producing contaminated aerosols, (b) number of susceptibles, (c) length of exposure, (d) ventilation rate, (e) settling rate of contaminated aerosols, and (f) the survival of pathogens attached to aerosols (16).

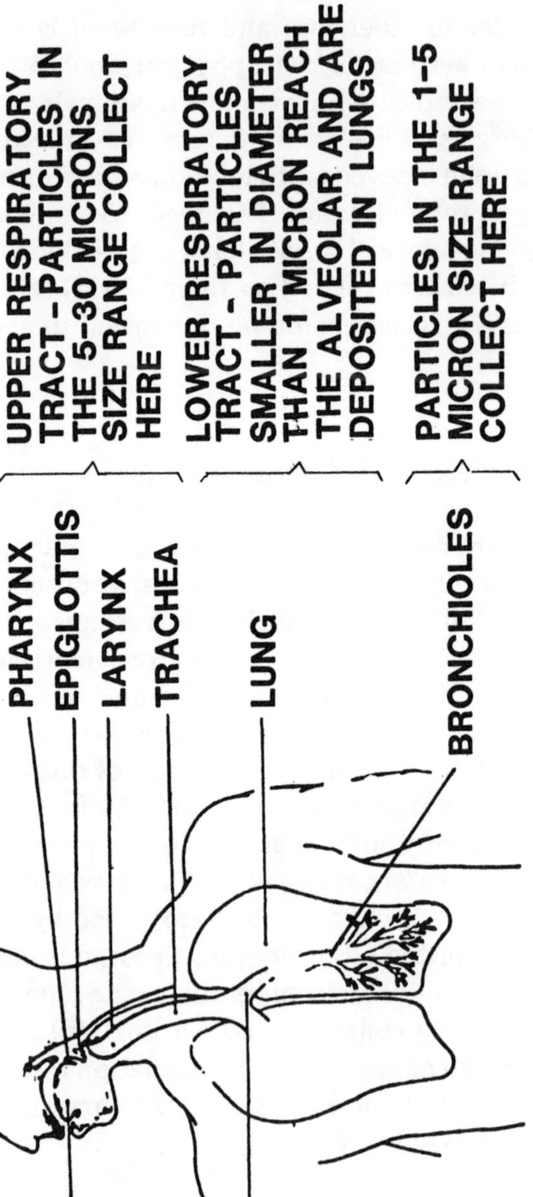

TONGUE
PHARYNX
EPIGLOTTIS
LARYNX
TRACHEA
BRONCHI
LUNG
BRONCHIOLES
UPPER RESPIRATORY TRACT - PARTICLES IN THE 5-30 MICRONS SIZE RANGE COLLECT HERE
LOWER RESPIRATORY TRACT - PARTICLES SMALLER IN DIAMETER THAN 1 MICRON REACH THE ALVEOLAR AND ARE DEPOSITED IN LUNGS
PARTICLES IN THE 1-5 MICRON SIZE RANGE COLLECT HERE

SOURCE : FIRE RESEARCH ON CELLULAR PLASTICS : THE FIRE REPORT OF THE PRODUCTS RESEARCH COMMITTEE, APRIL 1980

Figure 3-1. Deposition of Particulates in the Respiratory Tract.

Relative humidity can affect two of the six factors above: the settling rate of contaminated aerosols and the survival of airborne pathogens. Low relative humidity may increase the abundance of infective aerosols produced by coughing or exhaling. Rapid evaporation in dry air may decrease the diameter of some aerosols below the minimum limit required for a particulate to remain in suspension. At a relatively higher humidity level, the same aerosol may reach the floor before sufficient evaporation occurs (17). The amount of aerosols in a given volume of air partially depends on the settling rate, which is a function of air movement and the aerosol diameter. High settling rates reduce the abundance of aerosols which, in turn, reduce the probability of effective contact with aerosols contaminated with pathogenic substances.

To determine the effects of constant volume-constant temperature (CV-CT) systems on indoor air quality (IAQ) in contrast to variable-air-volume (VAV) systems, The Meckler Group tested respirable particulate concentrations in two Los Angeles office buildings. These concentration tests were used in a computer model based on the Skaret model (ASHRAE DC-83-09 No. 2).

The computer model was developed to analyze both one- and two-compartment models. It is based on an iterative numerical integration algorithm which permits evaluation of the total suspended particulate (TSP) net internal generation rate for a given distribution system by particulate size. The generated data base is utilized by this model to determine a set of constants for calculating the total internal generation rate of TSP for a given area.

The computer program simulates a two-compartment model as shown in Figure 3-2, and determines the TSP internal generation rate for each particle size range. Then, having generated a set of constants in the two-compartment model, the program simulates a one-compartment model by using a methodology developed by The Meckler Group. This enables the computer program to determine the ventilation efficiency, and predict future TSP concentrations by evaluating various changes in the compartment parameters.

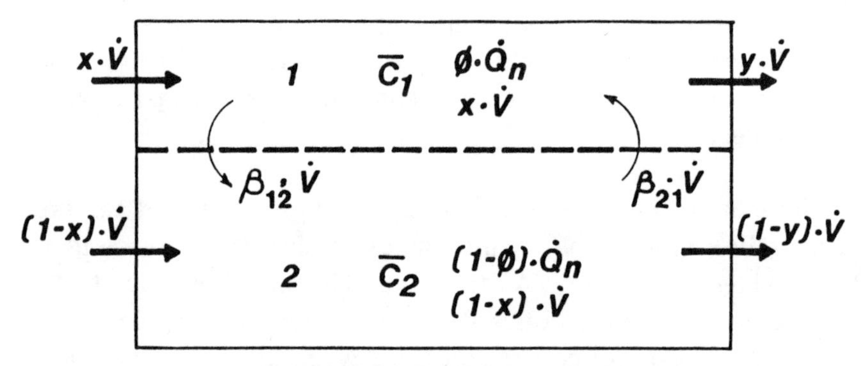

TWO-COMPARTMENT MODEL

NOMENCLATURE

α = Ratio of outside air to supply air flow rate
β = Interzonal mixing parameter
C_E = Concentration of contamination in exhaust air ($\mu g/m^3$)
$\overline{C}_1$ = Mean contaminant concentration in Zone 1 ($\mu g/m^3$)
$\overline{C}_2$ = Mean contaminant concentration in Zone 2 ($\mu g/m^3$)
C_O = Concentration of contamination in outside air ($\mu g/m^3$)
C_{CM} = "Perfect Mixing" contamination ($\mu g/m^3$)
C_S = Concentration of contaminants in supply air ($\mu g/m^3$)
C_Y = Concentration of contaminants in return air ($\mu g/m^3$)
ϵ = Filter effectiveness
F_r = VAV air flow reduction factor
n = Nominal air exchange rate ($\dot{V}/V$)
$\dot{N}$ = Total net internal production rate of contamination ($\mu g/m^3$)
ϕ = Relative production rate of contamination
$\dot{Q}$ = Contamination production rate ($\mu g/m^3$)
$\dot{Q}_n$ = Net Internal Production rate of contamination ($\mu g/m^3$)
V = Room volume (m^3)
$\dot{V}$ = Supply air flow rate (m^3/minute)
$\dot{V}_m$ = Outside air flow rate (m^3/minute)
$\dot{V}_R$ = Recirculate air flow rate (m^3/minute)
$\dot{V}$ = Exhaust air flow rate (m^3/minute)
E_v = Ventilation efficiency
X = Ratio of supply air rate for Zone 1 to supply air rate for Zone 2
Y = Ratio of exhaust air rate for Zone 1 to exhaust air rate for Zone 2

Figure 3-2. Two-Compartment Simulation Model.

Figures 3-3, 3-4, and 3-5 show the particle concentration vs. unit airflow for three particle size ranges tested: 0.3 to 1 micron, 1 to 3 microns, and 3 to 5 microns. The results suggest that VAV systems are subject to higher respirable concentration levels than CV-CT systems in tested air distribution systems with ceiling supply and return outlets.

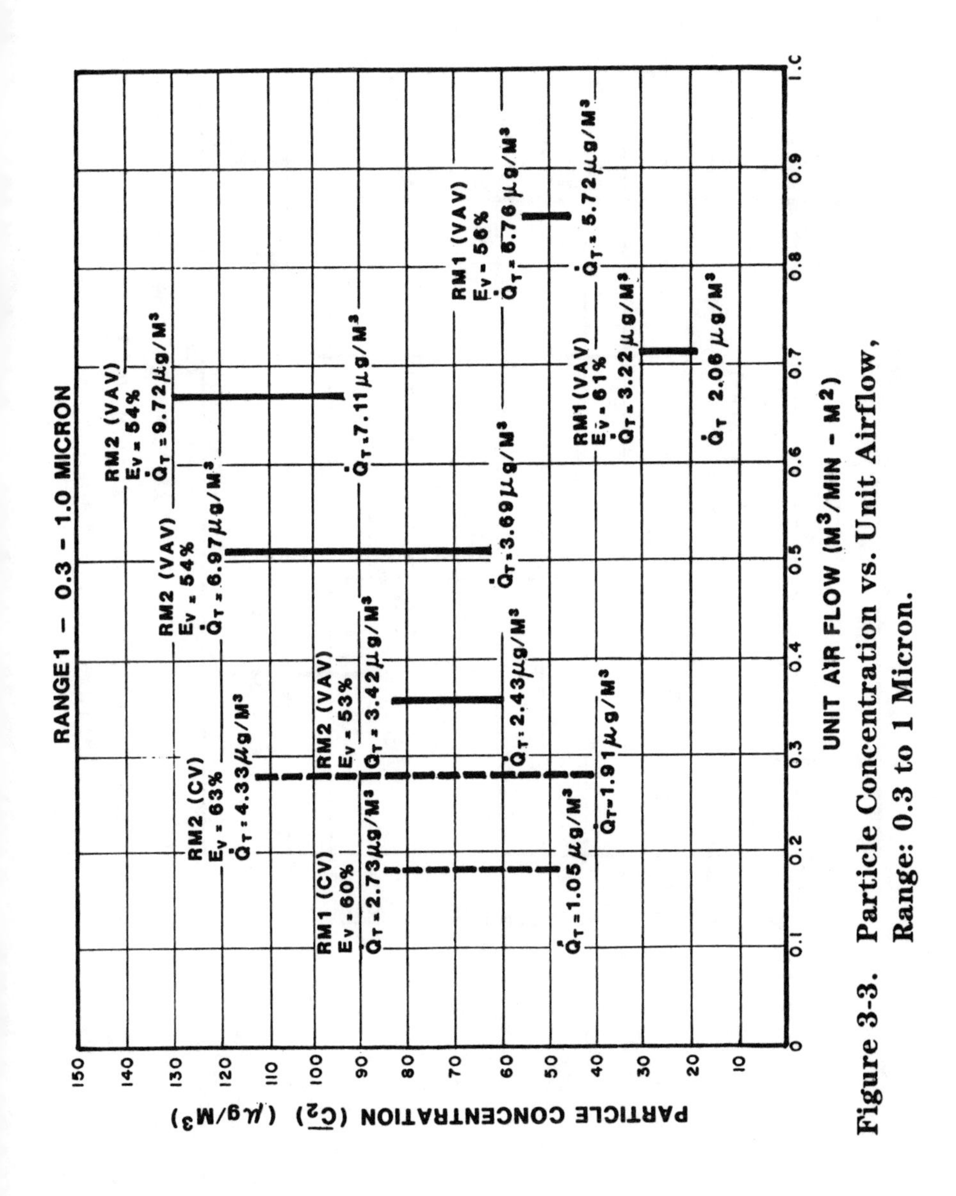

Figure 3-3. Particle Concentration vs. Unit Airflow, Range: 0.3 to 1 Micron.

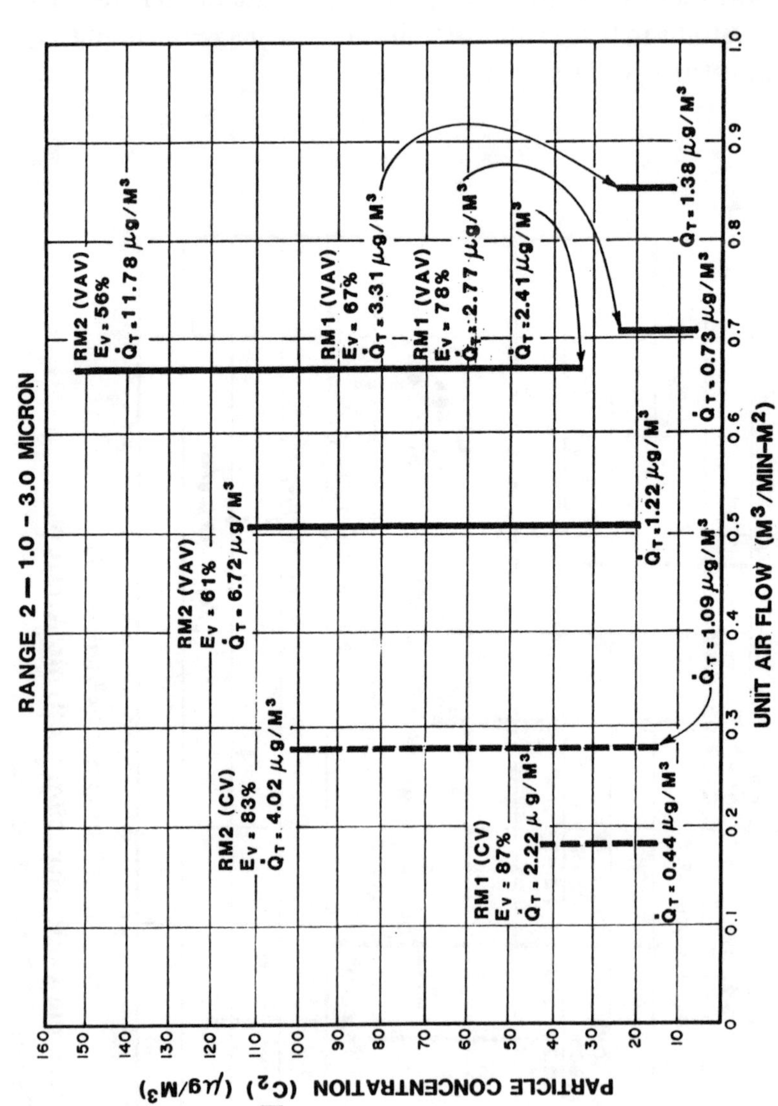

Figure 3-4. Particle Concentration vs. Unit Airflow, Range: 1 to 3 Microns.

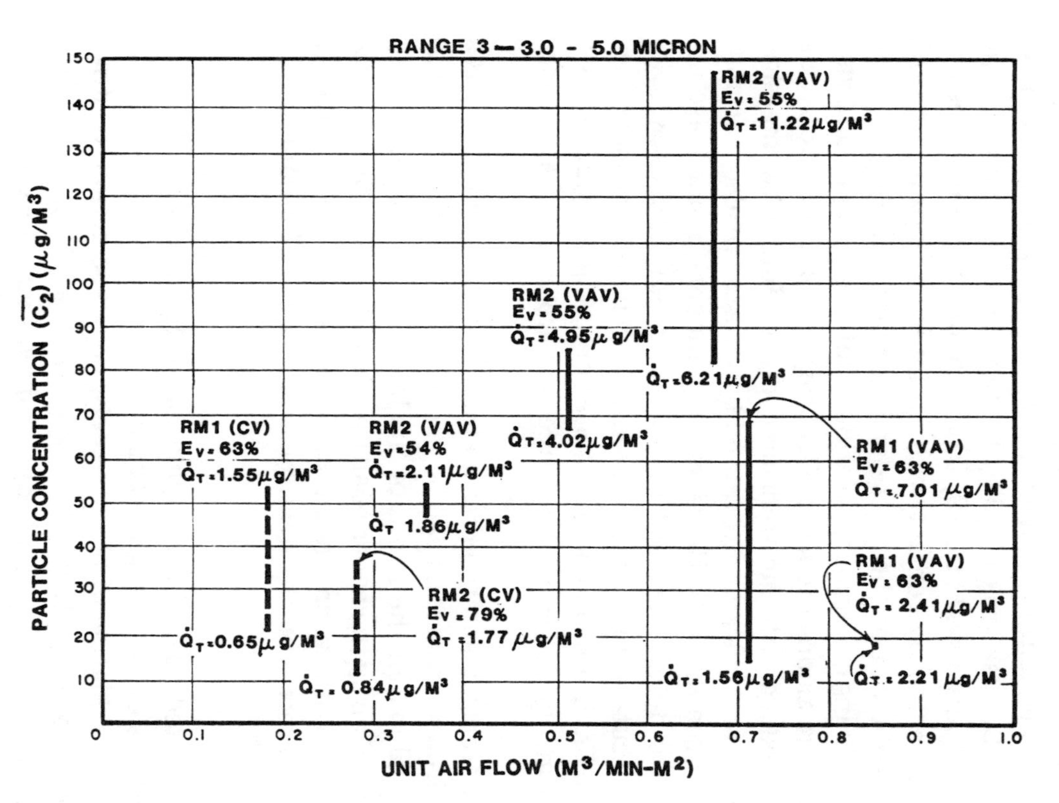

Figure 3-5. Particle Concentration vs. Unit Airflow, Range: 3 to 5 Microns.

The tests also suggest that the majority of the particles found were 2 to 5 microns in diameter. ETS particles are about 1 micron in diameter.

Psittacosis, Q-fever, brucellosis, pneumonic tularemia, pneumonic plaque, anthrax, tuberculosis, whooping cough and Legionnaires Disease are caused by bacteria and are believed to be airborne transmitted (18, 19, 20). Although these diseases are also contracted by contact with an infected person, airborne transmission can be the most dominant mode of transfer. Less crowded living conditions, isolation of infected people, and vaccination are among the most effective control techniques available to reduce airborne transmission (21). Influenza, measles, smallpox, and chicken pox caused by viruses are also transmitted airborne (20, 22). However, experimental studies are necessary to determine the transmission route for most viruses causing acute respiratory infections. Flush toilets, ice machines, hospital environments, and factories processing organic materials are among the potential sources to harbor bacteria aerosols.

Fomite transmission requires an intermediary fomite such as a cup. Viruses or bacteria can be deposited on a fomite by sneezing, coughing, or the hands of the infected person where they are picked up by the hands of an uninfected person. Direct transmission occurs when an infected person touches, coughs or sneezes directly on the eyes, nose, or mouth of an uninfected person.

An estimated 10% of the population suffer from allergies (23). The effects of the allergic reactions on a particular body tissue include the dilation of blood vessels, mucus secretion, contraction of the bronchial smooth muscles, and cellular inflammation (21). Reduced airflow rates have an important effect on the concentrations of airborne allergenic agents. They tend to provide a favorable medium for molds, housedust, fungi, etc. Fungi is one of the major causes of allergy. Ceiling tiles in offices can be a common source of fungous contamination, as the tiles may be directly exposed to moisture when the air-conditioning system is in use. In addition, damp organic materials such as leather, cotton, paper, furniture stuffing, and carpets can be contaminated with fungi (24). In addition, interiors used specifically for processing and handling of biological materials are expected to have substantially higher fungous contamination.

Pollens from trees, grass, and flowers have been associated with allergic rhinitis and asthma (21). The typical symptoms are stuffy or runny nose in allergic rhinitis and breathing difficulty in asthma. Although pollens are generally too large to penetrate beyond the trachea, the fragments of pollens may very well penetrate lower airways. Also housedust is determined to be one of the major causes of allergic rhinites and asthma in humid and mild climates. Mites are the most important cause of housedust allergies. They are easily found in beds, mattresses and pillows and, require high humidity (21).

References

1. Ing, R.; Falk, H.; Lemen, R.; Carson, G.; Kelter, A.; Sarn, J.; and Gray, M. J. of Pediatrics, 99, p. 409, 1981.

2. Silver, K.Z. Ann. NY Acad. Sci., 330, p. 777, 1979.

3. Nicholson, W.J.; Swoszowski, E.J.; Rohl, A.N.; Todaro, J.M.; and Adams, A. Ann. NY Acad. Sci., 330, p. 587, 1979.

4. Sawyer, R.N. Ann. NY Acad. Sci., 330, p. 579, 1979.

5. Sterling, E.M. "Asbestos Contamination of Public and Commercial Buildings." *BOMA BC News*, Guest Editorial, 1987.

6. Sterling T.D. and Arundel, A. J. Environ. Sci. and Health, C2(2), pp. 185-230, 1984.

7. McDonald, J.C.; Liddell, F.D.K.; Gibbs, G.W.; Eyssen, G.E.; and McDonald, A.D. Br. J. Ind. Med., 37, p. 11, 1980.

8. Selikoff, I.J. and Seidman, H. Cancer, 47, p. 1469, 1981.

9. Newhouse, M.L. and Berry, G. Ann. NY Acad. Sci., 330, p. 53, 1979.

10. Anderson, H.A.; Lilis, R.; Daum, S.M.; and Selikoff, I.J. Ann. NY Acad. Sci., 330, p. 387, 1979.

11. Fischbein, A. and Rohl, A.N. JAMA, 252, p. 86, 1984.

12. Newhouse, M.L. and Thompson, H. Br. J. Ind. Med., 22, p. 261, 1972.

13. Newhouse, M. Seminars in Oncology, 8, p. 250, 1981.

14. Sterling, T.D. and Kobayashi, D.M. Environ. Research, 13, p. 1, 1977.

15. Arundel, A.V. and Sterling, E.M. "Review of the Effect of Relative Humidity on the Transmission of Infectious Diseases." 80th annual meeting of APCA, New York, NY, 1987.

16. Couch, R.B. "Viruses and Indoor Air Pollution." Bull. NY, *Acad. Med.*, 57, pp. 907-921, 1981.

17. Smith, F.B. "Atmospheric Factors Affecting Transmission of Infections." *Practitioner*, U.K., 227, pp. 1667-1677, 1983.

18. Fraser, D.W. "Legionellosis Evidence of Airborne Transmission." *Ann. NY Acad. Sci.*, p. 353, 1980.

19. Kaufmann, A.F.; Fox, M.D.; Boyce, J.M.; Anderson, D.C.; Potter, M.E.; Martone, W.J.; and Potter, C.M. "Airborne Spread of Brucellosis." *Ann. NY Acad. Sci.*, 355, p. 105, 1980.

20. Langmuir, A.D. "Changing Concepts of Airborne Infection of Acute Contagious Diseases: A Reconsideration of Classic Epidemiological Theories." *Ann. NY Acad. Sci.*, 353, p. 35, 1980.

21. Turiel, I. Indoor Air Quality and Human Health, 1985.

22. Gregg, M.B. "The Epidemiology of Influenza in Humans." *Ann. NY Acad. Sci.*, 353, p. 45, 1980.

23. Pepys, J. "Clinical Therapeutic Significance of Patterns of Allergic Reactions of the Lungs to Extrinsic Agents." *Am. Rev. Resp. Dis.*, 1, p. 116, 1977.

24. National Academy of Sciences. Indoor Pollutants, National Academy Press, Washington, DC, 1981.

4. MAJOR COMBUSTION BYPRODUCTS

Larry C. Holcomb, Ph.D.
President, Holcomb Environmental Services
Olivet, Michigan

and

Elia M. Sterling
Director of Building Research, Theodor D. Sterling
and Associates, Ltd.
Vancouver, B.C., Canada

Page

4.1 Introduction

Byproducts of combustion indoors can come from several different sources. The four major sources that have the largest contribution to indoor air contamination are: (a) gas stoves, (b) unvented space heaters, (c) wood stoves and fireplaces, and (d) tobacco smoke. Nicotine, one of the major combustion byproducts of tobacco smoke, has been the subject of considerable research and is discussed in detail in Chapter 5, Part 1, separately. The major combustion byproducts include nitrogen dioxide (NO_2), carbon monoxide (NO), sulfur dioxide (SO_2), and particulates. Although carbon dioxide (CO_2) is a combustion byproduct, it is considered to be relatively nontoxic. However, it is a very good indicator of indoor air quality (IAQ) and the adequacy of ventilation. There are several other significant outdoor sources of combustion byproducts which can affect IAQ, such as vehicle exhaust, fossil fuel burning plants, reentrainment of improperly vented combustion fumes from indoors, etc. (1).

Gas stoves and unvented space heaters have been associated with high concentrations of NO_2 and CO in indoor environments. Carbon monoxide concentrations have been shown to exceed 12 ppm in homes where gas is used for cooking. In situations where gas stoves are improperly used for heating, CO concentrations may be as high as 25 ppm to 50 ppm (2). Several studies (1,3,4) have associated the use of space heaters and gas stoves (particularly with pilot lights) with increased concentrations of NO_2. According to one study (3), the average concentration of NO_2 when cooking with a gas stove was 168 ppb in comparison to 22 ppb in homes that used electricity for cooking.

Lately, heating with wood has become an increasingly attractive option. At the present time, over 11 million wood stoves are estimated to be in use (2). Since wood burning takes place in a relatively inefficient environment for combustion, several byproducts are produced. The major byproducts are CO, respirable particulates, and polycyclic aromatic hydrocarbons (PAHs). The level of indoor air contamination due to the use of wood stoves is determined primarily by its airtightness. New and airtight wood stoves seem to have a positive effect on indoor air contamination. The use of older and non-airtight stoves has produced concentrations of 14 ppm of CO, 200 $\mu g/m^3$ to 1900 $\mu g/m^3$ of respirable particulates, and PAH concentrations that greatly exceeded outdoor levels (2).

4.2 Carbon Monoxide

Carbon monoxide is a colorless, odorless and tasteless gas. It is a byproduct of the combustion of carbon-containing materials in an oxygen deficient environment. The toxic effects of CO are believed to be due to the fact that it combines with the hemoglobin in the blood to form carboxyhemoglobin (COHb). This interferes with the ability of the blood to carry oxygen and results in oxygen starvation to the body tissue. The parts of the body tissue with the highest oxygen demand such as the brain and heart are most strongly affected. There is some evidence to suggest that CO may also affect the ability of myoglobin and cytochrome oxidase to handle oxygen, thus increasing its ability to deprive tissues of oxygen. The amount of COHb formed in the body depends on the concentration of CO in the air. Measured relationships between the CO concentrations in the air and COHb levels in the blood of nonsmokers are as follows:

15 ppm CO - 2.4% COHb
50 ppm CO - 7.1% COHb
75 ppm CO - 10.9% COHb

COHb levels in the blood of smokers can range from 5% to 10% higher than those in nonsmokers at the same CO concentrations (5).

The health effects due to high-level exposure to CO can range from headaches and dizziness to nausea and vomiting and to coma and death. The health effects due to low-level exposure to CO (COHb of 2% to 5%) are not well defined. In young and healthy adults, decreased oxygen intake ability and work capacity have been measured at COHb levels as low as 5%. Patients suffering from angina have been shown to be affected by COHb levels as low as 2.9%. Neurobehavioral functions (as measured by sensory and time discrimination, sensorimotor performance, and sleep activity) have been shown to be affected at COHb levels of 5%. Due to the nature of tests and their shortcomings, it is impossible to determine whether lower COHb levels affect neurobehavioral functions (5). Other health effects due to low-level exposure to CO such as slowed infant development, increased risk of sudden infant death syndrome, and effects on fibrinolytic activity have been suggested but not confirmed.

Carbon monoxide is one of the major indoor air pollutants emitted during the operation of combustion appliances. It has always been assumed that only a small amount of CO will be produced by a gas stove and, this amount will be dissipated by the ventilation system and a combination of a fan and hood over the stove. However, cooking may substantially increase the level of CO. The air supply may be progressively diminished when more than one burner is used, and air supply may be partially cut off by vessels placed over the gas flame.

In one study (6), nine kitchens with gas stoves were randomly selected. In one of these kitchens, a series of experiments was conducted to determine the growth and dissipation of CO under normal cooking conditions and ventilation types. Figure 4-1 shows the increase in CO when using 1 to 4 covered and uncovered burners. The rate of increase is sharply linear during cooking and depends on the number of burners used. As can be seen, placing a pan over the flame substantially increases the rate of CO production in all four cases. This is due to the fact that air is cut off to some extent and incomplete combustion takes place. Once the CO spreads, it dissipates slowly depending on the type of ventilation. Figure 4-2 shows CO concentrations in the kitchen, dining room (next to kitchen), and the living room (next to dining room) for 90 minutes after the stoves had been turned off. Figure 4-3 shows the disappearance of CO from the kitchen under various ventilation conditions. The use of a hood and fan over the stove or opening windows (without cross ventilation) results in moderate decrease in CO. On the other hand, effective cross ventilation will reduce the CO concentrations rapidly, as shown in Figure 4-3.

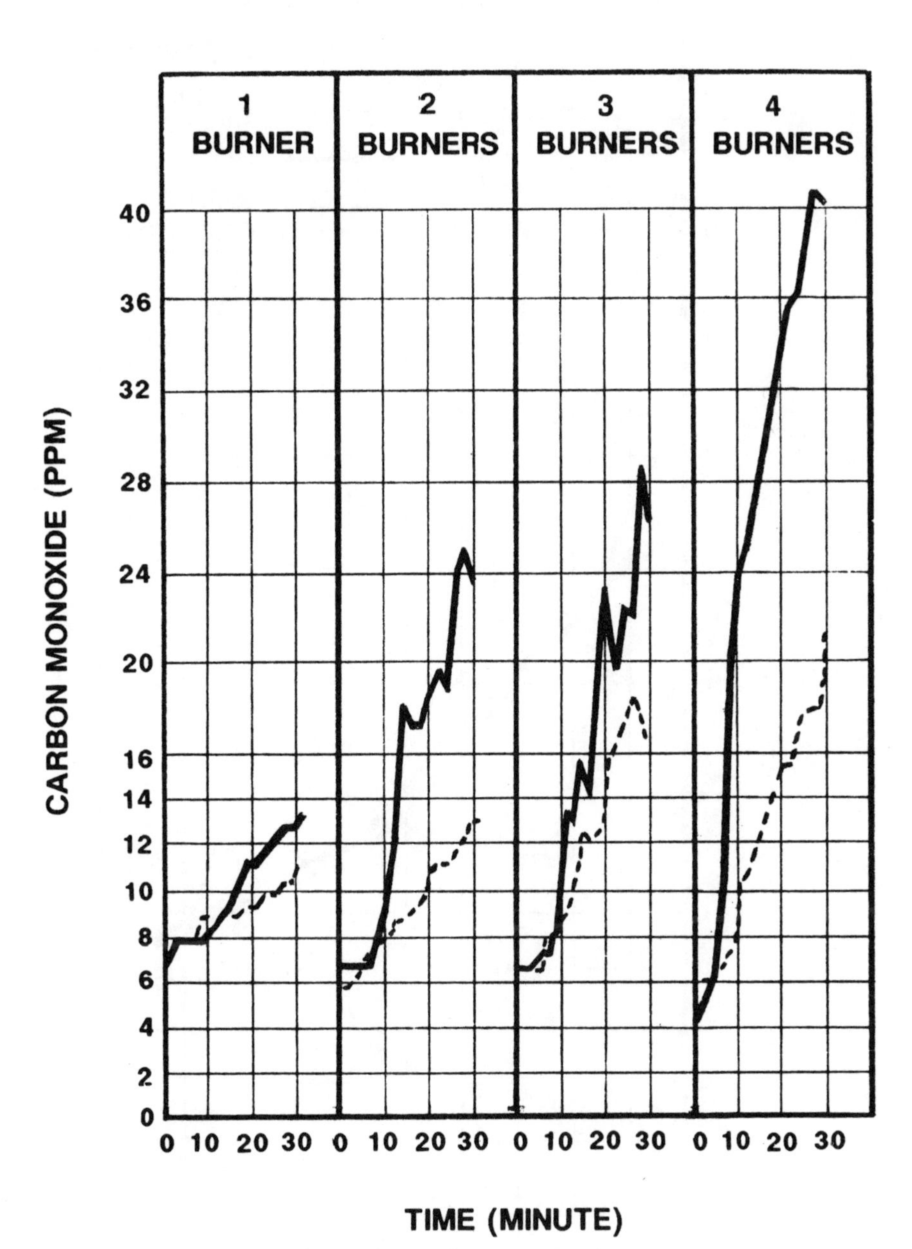

Figure 4-1. Increase in Carbon Monoxide (1 to 4 Covered and Uncovered Burners).

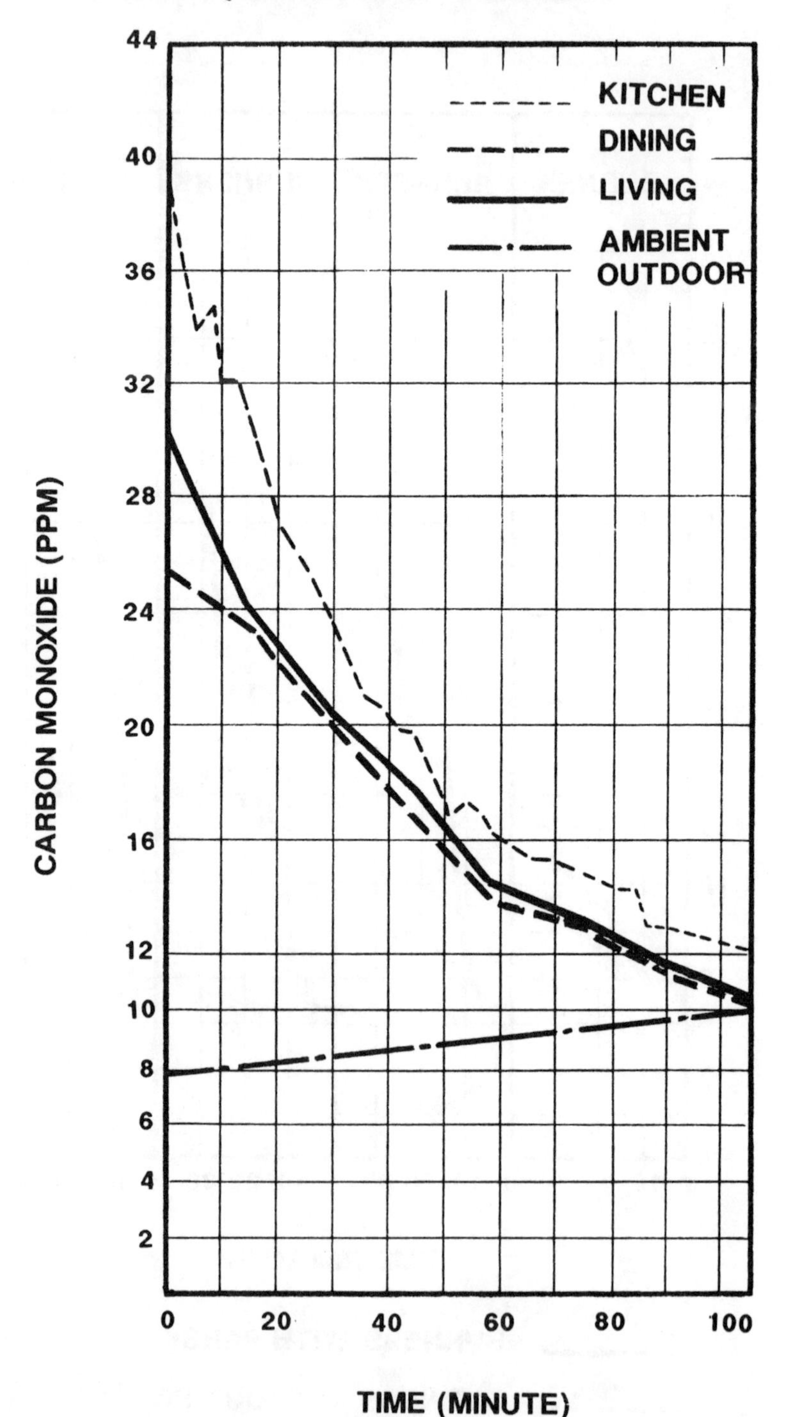

Figure 4-2. Carbon Monoxide Levels in Kitchen, Dining Room and Living Room.

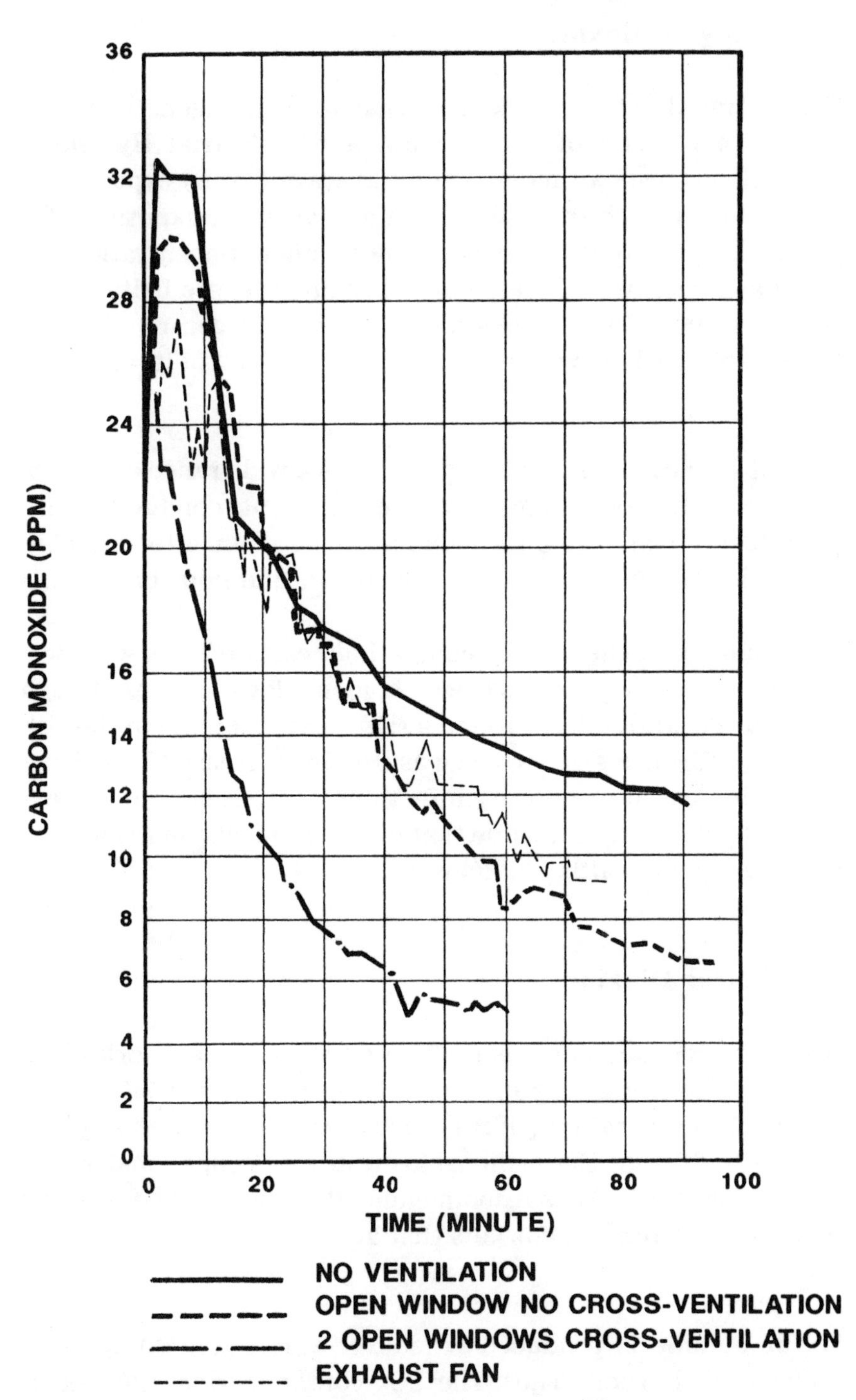

Figure 4-3. Decrease of Carbon Monoxide in Kitchen Under Various Conditions and Ventilation.

4.3 Nitrogen Dioxide

Concentrated NO_2 is a dark brown gas with a pungent odor. The American Council of Government and Industrial Hygienists (ACGIH) suggests a time-weighted average (TWA) exposure of 3 ppm for an eight-hour day with a short-term exposure of 5 ppm. The physiological effects due to high concentrations of NO_2 have been well documented. It causes severe irritation to the eyes and other mucous membranes. Nitrogen dioxide has been shown to damage the lungs directly because of its oxidizing ability.

Animal studies have shown that NO_2 has a wide range of effects on the respiratory system. Several experiments conducted with volunteer human beings have indicated minimum effects at concentrations above 1 ppm in exposures of one hour or longer.

The majority of the studies comparing health effects with residential NO_2 exposures have been epidemiological. Even though some of the studies have found a significant increase in the risk of adverse health effects, as measured by decreased lung function or increased respiratory infections, there is not enough evidence to positively conclude that exposure to NO_2 in residences causes serious health problems.

4.4 Sulfur Dioxide

Another important combustion byproduct is SO_2, a colorless gas with a suffocating type of odor. The Occupational Safety and Health Administration (OSHA) regulates SO_2 in the workplace at a TWA of 5 ppm. The National Institute for Occupational Safety and Health (NIOSH) has recommended that this level be lowered to 0.5 ppm to protect workers that are particularly sensitive to SO_2 (7).

Sulfur dioxide is produced whenever sulfur-containing compounds are burned. Thus, the SO_2 contamination may result from several combustion sources. Since SO_2 is highly soluble in water, it is readily absorbed by the mucous membranes of the respiratory system. Upon inhalation, it is dissolved by the moisture in the mucous membranes and forms sulfurous acid, sulfuric acid, and bisulfate ions. During normal nasal respiration patterns, SO_2 is absorbed primarily by the nasal tissues and only 1% to 5% reaches the lower respiratory tract. However, during oronasal and oral breathing (during exercising or in the presence of some respiratory infections), significant quantities of SO_2 can penetrate to the lower respiratory tract even at low ambient concentrations.

The primary effects of SO_2 exposure occur in the respiratory tract. It causes significant bronchoconstriction in asthmatics at concentrations below 1 ppm. For healthy human beings, this effect begins at concentrations of approximately 5 ppm (8). The constriction associated with SO_2 typically develops a few minutes following the exposure and goes away within minutes after the exposure stops. The intensity of constriction is a function of the number of SO_2 molecules reaching the lower airways per unit time, not the total dose received. Its effect does not increase with exposure time.

Several studies that measured excess death rates in cities with high air pollution have concluded that the SO_2 concentrations over 0.4 ppm when associated with high particulate levels (greater than 1000 $\mu g/m^3$) and high humidity result in markedly increased mortality (9). This effect has been most pronounced when the levels of contaminants remained elevated for several days.

4.5 Carbon Dioxide

Carbon dioxide is considered to be relatively nontoxic. It is an excellent indicator of the adequacy of ventilation and may be used to determine whether other indoor air contaminants accumulate in a space. It controls the respiratory rates in a person. If the level of CO_2 is allowed to build-up, people soon feel as if there is not enough air in the room; their respiratory rate accelerates in an attempt to compensate for it.

The OSHA and ACGIH both have set a permissible exposure level (PEL) for CO_2 at 5000 ppm for an eight-hour TWA. Outdoor concentrations of CO_2 generally vary between 350 ppm and 400 ppm. When indoor concentrations exceed 1000 ppm, the air in a room becomes stuffy. Carbon dioxide is generated by people at a rate of about 0.3 L/m/person. If at any time the ventilation rate in a room fails to keep up with this production rate, the concentration of CO_2 will increase. ASHRAE Standard 62-1989 recommends a maximum CO_2 concentration of 1000 ppm, which is a ventilation requirement of 15 cfm/person.

References

1. Petreas, M.; Liu, K.S.; Chang, B.H.; Hayward, S.B.; and Sexton, K. "A Survey of Nitrogen Dioxide Levels Measured Inside Mobile Homes." *JAPCA*, 38, pp. 647-651, 1988.

2. Samet, J.M.; Marbury, M.C.; and Spengler, J.D. "Health Effects and Sources of Indoor Air Pollution." Pt. I. *Am. Rev. Respir. Dis.*, 136, pp. 1486-1508, 1987.

3. Marbury, M.C.; Harlos, D.P.; Samet, J.M.; and Spengler, J.D. "Indoor Residential NO_2 Concentrations in Albuquerque, New Mexico." *JAPCA*, 38, pp. 392-398, 1988.

4. Traynor, G.W.; Apte, M.B.; Carruthers, A.R.; Dilworth, J.F.; Prill, R.J.; Grimsrud, D.T.; and Turk, G.H. "The Effects of Infiltration and Insulation on the Source Strengths and Indoor Air Pollution from Combustion Space Heating Appliances." *JAPCA*, 38, pp. 1011-1015, 1988.

5. Environmental Protection Agency. "Revised Evaluation of Health Effects Associated with Carbon Monoxide Exposure: An Addendum to the 1979 EPA Air Quality Criteria Document for Carbon Monoxide." Environmental Criteria

and Assessment Office, Office of Health and Environmental Assessment, NC, 1984.

6. Sterling, T.D. and Sterling, E.M. ''Carbon Monoxide Levels in Kitchens and Homes with Gas Cookers.'' *APCA*, Vol. 29, No. 3, 1979.

7. Sittig, M. Handbook of Toxic and Hazardous Chemicals and Carcinogens. 2nd Ed., Noyes Publications, NJ, 1985.

8. Environmental Protection Agency. ''Air Quality Criteria for Particulate Matter and Sulfur Oxides.'' Vol. 1. Environmental Criteria and Assessment Office, Office of Health and Environmental Assessment, NC, 1982.

9. Environmental Protection Agency. ''Second Addendum to Air Quality Criteria for Particulate Matter and Sulfur Oxides.'' Environmental Criteria and Assessment Office, Office of Health and Environmental Assessment, NC, 1986.

5. OTHER INDOOR AIR POLLUTANTS

Larry C. Holcomb, Ph.D.
President, Holcomb Environmental Services
Olivet, Michigan

and

Elia M. Sterling
Director of Building Research, Theodor D. Sterling
and Associates, Ltd.
Vancouver, B.C., Canada

5.1 Introduction

There are several thousands of different household products which pose a serious threat to indoor air quality (IAQ). These products include cleaners, detergents, paints, air fresheners and disinfectants, furniture polishes, insecticides, etc. In this chapter, we will concentrate on the volatile organic compounds (VOCs) found indoors and environmental tobacco smoke (ETS).

5.2 Volatile Organic Compounds

VOCs consist of a large and diverse group of substances that evaporate into the atmosphere at room temperature. They are generally used as solvents in the formulation or manufacturing of consumer products. Presently, more than 350 VOCs have been identified indoors with concentrations exceeding 0.001 ppm (1). VOCs are part of almost all materials and products such as construction materials, furnishings, combustion fuels, consumer products, and pesticides.

VOCs fall into two major groups: petroleum-based solvents and chlorinated solvents. The petroleum-based solvents are generally found in products such as paints, stains, adhesives, and caulks. Chlorinated solvents are common ingredients of shoe polishes, water repellents, epoxy paint sprays, paint removers, and dry cleaning compounds. Some of these chlorinated solvents are carbon tetrachloride, dichloroethane, trichloroethane, dichloroethylene, trichloroethylene, tetrachloroethylene, etc. Organic compounds are also commonly used as propellants in several types of aerosol products (2). Table 5-1 shows the types of VOCs commonly found indoors and their sources. In addition, many pesticides are commonly used in homes, gardens, or lawns to kill various insects. These include chlordane, heptachlor, malathion, diazinon, dursban, ronnel, and dichlorvos.

The indoor VOC concentrations are often five to ten times higher than outdoor concentrations. This implies that there are significant indoor sources. Identification of individual sources of these VOCs has been very difficult. Although several different VOCs have been identified indoors, their concentrations tend to be below the regulated levels in occupational environments. There is still concern over the potential health effects of this very low-level exposure. Exposures are often multiple, complaint symptoms are common to many compounds, and the population is very diverse. In general, VOCs are lipid soluble and easily absorbed through the lungs. Their ability to readily cross the blood/brain barrier may induce central nervous system (CNS) depression, a possible causal factor in drowsiness, fatigue and general malaise (1). Alcohols, aromatic hydrocarbons and aldehydes can irritate mucous membranes.

Table 5-1. Common Volatile Organic Compounds and Indoor Sources.

Pollutant Type	Example	Indoor Source
Aliphatic hydrocarbons	Propane, butane hexane, limonene	Cooking and heating fuels, aerosol propellants, cleaning compounds, refrigerants, lubricants, flavoring agents, perfume base
Halogenated hydrocarbons	Methyl chloroform, methylene chloride,	Aerosol propellants, fumigants, pesticides, refrigerants, degreasing, dewaxing and dry cleaning solvents
Aromatic hydrocarbons	Benzene, toluene, xylenes	Paints, varnishes, glues, enamels, lacquers, household cleaners
Alcohols	Ethanol, methanol	Window cleaners, paints, thinners, cosmetics, adhesives, human breath
Ketones	Acetone	Lacquers, varnishes, polish removers, adhesives
Aldehydes	Formaldehyde, nonanal	Fungicides, germicides, disinfectants, artificial and permanent-press textiles, urea-formaldehyde-foam insulation (UFFI), paper, particle boards, cosmetics, flavoring agents

One theoretical study (3) on carcinogenicity of several VOCs commonly found indoors, reported an increased risk of cancer over a lifetime exposure from 0.2% to 3%. Although this study was theoretical in nature and based on several assumptions, it provided an understanding of the risks involved in low-level exposures. With the increasing number of studies, it may be possible to begin to isolate the VOCs commonly found indoors, identify their potential sources, determine possible health hazards, and take precautionary measures.

5.3 Environmental Tobacco Smoke

Some recent studies have reported increased relative risk to non-smokers due to ETS exposure. It is the most visible indoor source of combustion byproducts. Therefore, recent attempts to clean the air in modern, sealed office buildings appear to be focused on ETS (4).

Burning of tobacco products produces a wide range of combustion byproducts such as carbon monoxide, particulates, nitrogen oxides, aromatic hydrocarbons, acrolein, aldehydes, nicotine, nitrosamines, hydrogen cyanide, and ketones (5, 6, 7). The combination of these byproducts is called ETS. ETS is a mixture of mainstream smoke exhaled by active smokers and sidestream smoke which comes off the burning end of a cigarette. When these two mixtures combine in the air they become highly diluted and undergo several chemical changes known as aging to form ETS (8). Because ETS is a very dynamic substance containing several components, monitoring ETS indoors has been very difficult. Another factor that has complicated the study of ETS indoors is that most compounds (except for nicotine) have other significant sources. Usually one or more components of the mixture is quantified and related back to ETS as a whole. The two most commonly measured components are the respirable suspended particles (RSP) and nicotine.

Even with accurate measurements of RSP in a room, it is estimated that, in most smoking environments, approximately one half of the RSP is contributed by ETS. The rest is produced by other sources. Current estimates of particulate concentrations generated by ETS indoors range from 20 $\mu g/m^3$ in residences to 260 $\mu g/m^3$ in restaurants and bars (9).

Nicotine tends to partition with the gaseous phase rather than the particulate phase of ETS. Consequently, its concentration in a room decreases much faster than that of particulate phase components. Measuring nicotine in an environment may tend to underestimate the particulate components of ETS. Nicotine concentrations in smoking areas with reasonable ventilation range from 1.6 μ g/m^3 to 53 μ g/m^3 (10).

Nicotine and its metabolic derivative, cotinine, have been measured in the body fluids of people exposed to ETS. Cotinine has a half-life of 16 to 19 hours. As a result, the cotinine levels depend not only on the quantity of ETS a person is exposed to, but the length of exposure time. This has made it difficult to back-calculate from cotinine levels to ETS concentrations a person has been exposed to (11). Currently, cotinine measurements are mainly used in health studies to differentiate between the smokers and nonsmokers.

Some of the combustion byproducts in public buildings have been widely monitored. Although in offices with proper ventilation and occupancy the concentrations of pollutants seem to vary little between smoking and non-smoking areas, this should not be construed as a rule. Concentrations of ETS in bars, restaurants, nightclubs, lobbies, and waiting rooms when combined with inadequate or no ventilation may be very high.

IAQ in commercial buildings has become a serious concern among the building owners and their clients. Unfortunately, complaints about temperature, ventilation, and odor have become commonplace. In some cases, the complaints have become serious enough to close down buildings leading to loss of revenues and costly litigations. Building operators are now desperately searching for workable solutions to this problem.

One solution that has gained wide support, by both the government and private corporations, is the regulation of smoking in buildings. Four options are available to regulate smoking in buildings (10):

- Prohibit smoking,
- Restrict smoking to those areas that are ventilated separately,

- Restrict smoking to those areas that are not ventilated separately, and

- Provide a framework in which an adjustment between the smokers and nonsmokers may be accomplished without designating a separate location for smokers.

The third option above seems to be the most frequently adopted option. Since most buildings do not have separate ventilation for different locations, providing such ventilation may be very costly and often physically impossible.

In one study (10), the concentrations of nicotine, RSP, carbon monoxide and carbon dioxide were measured in offices under various conditions of smoking regulations: smoking prohibited; smoking prohibited areas receiving recirculated air from designated smoking areas; and smoking and non-smoking sections of these designated smoking areas. Tables 5-2 and 5-3 tabulate the results of the above-mentioned study in: (a) two cafeterias, each with smoking and non-smoking areas; (b) four non-smoking floors, receiving recirculated air from a ventilation system common to one of the cafeterias; and (c) two non-smoking offices with independent ventilation systems which, therefore, did not receive recirculated air from the designated smoking areas.

Detection of nicotine in non-smoking office areas that received recirculated air from the designated smoking areas required sampling times of four hours or more. Nicotine levels in such offices were approximately 1.0 $\mu g/m^3$. RSP, carbon monoxide, and carbon dioxide concentrations were approximately the same in these offices as compared to non-smoking offices not receiving recirculated air from the smoking areas. Based on these findings, it can be seen that providing a designated smoking area appears to be effective in eliminating most ETS from the rest of the office space even if the designated smoking area is not separately ventilated. If a separate smoking area is to be designated, this area should be sufficiently large to prevent overcrowding.

Table 5-2. Comparison of Environmental Tobacco Smoke Air Quality Parameters in Non-Smoking Work Areas and Designated Smoking Areas.

Area		RSP[A] (μ g/m^3)	CO (ppm)	CO_2 (ppm)	Nicotine (μ g/m^3)	Persons /10m^2	Cigarettes /h/10 m^2
Smoking areas	Mean	70	3.9	690	14	1.8	1.2
of Cafeterias A	Range	23-129	1.1-11.4	450-1000	<1.6-43.7	0.79-3.42	0.53-1.67
& B combined	Median	74	2.5	650	11	1.6	1.2
Non-smoking ar-	Mean	32	2.6	560	6.2	1.7	
eas of Cafeterias	Range	15-57	1.2-4.5	400-700	<1.6-10.9	0.76-2.5	NA
A & B Combined	Median	26	2.4	580	7.9	1.7	
Non-smoking	Mean	6	1.8	490		0.73	
office area,	Range	4-11	1.3-2.3	400-580	B	0.28-1.9	NA
Building A	Median	6	1.7	500		0.46	
Non-smoking	Mean	7	1.35	450		0.9	
office area,	Range	6-8	1.3-1.4	400-500	B	0.53-1.28	NA
Building B	Median	7	1.35	450		0.9	

A: Mean outdoor RSPs were 10 μ g/m^3.

NA: Not applicable.

B: Refer to Table 5-3.

Table 5-3. Concentrations of Nicotine, Respirable Suspended Particles, Carbon Monoxide and Carbon Dioxide in Eight Locations in a Non-Smoking Office Receiving Recirculated Air from a Designated Smoking Area, and in Two Locations without Recirculation[A].

	Location	Sample Time (hr)	Nicotine ($\mu g/m^3$)	RSP ($\mu g/m^3$)	CO (ppm)	CO_2 (ppm)	Persons /10 m^2
Recirculated air	1	2	<0.8	6	1.7	580	0.50
	2	2	<0.8	5	2.3	500	0.42
	3	2	<0.8	5	1.3	400	1.90
	4	2	<0.8	4	2.0	500	0.38
	5	4	<0.4	11	2.2	550	0.39
	6	4	<0.4	5	1.7	450	0.28
	7	4	1.0	6	1.7	500	1.02
	8	8	0.8	6	1.6	450	0.96
No recirculated air	9	B	B	8	1.4	400	0.53
	10	B	B	6	1.3	500	1.28

A: Sample time for nicotine: 2 to 8 hours.
B: Nicotine was not measured because these offices could not receive ETS from any source.

Major reviews on ETS and its health effects have been done by the U.S. Surgeon General (12) and the National Academy of Sciences (8). These reports can be broken down into three major groups of health effects: health effects in children, cancer in adults, and non-cancer diseases in adults. To date, there are inherent weaknesses in all of the studies on ETS and its possible health effects. All studies use indirect measures of exposure (questionnaires) and have problems with confounding factors. Virtually all of them employ epidemiology studies; these studies cannot show any cause-and-effect relationship. Although scientific controversy on the subject continues, there is still no persuasive evidence that ETS is the cause of any disease in nonsmokers.

Since 1971, the National Institute for Occupational Safety and Health (NIOSH) has conducted over 350 investigations of buildings with health and comfort problems. The findings from 203 of these investigations through 1983 were reviewed and tabulated by the Health Hazard Evaluation Branch of NIOSH (13). Table 5-4 lists the apparent causes of the IAQ problems for these investigations. Cigarette smoking was identified as a contributing factor in only 2% of the investigations. By far, the most apparent problem was inadequate ventilation (48.3%). Ventilation becomes inadequate when the amount of outside fresh air is not sufficient enough to dilute the concentrations of indoor air contaminants. This is usually done to conserve energy and reduce building operation expenses.

Table 5-4. Suspected Causes of Indoor Air Quality Problems (National Institute for Occupational Safety and Health, Health Hazard Evaluation Branch (13)).

Suspected Cause	No. of Investigations	Percent of Total
Contamination (inside)	36	17.7
Contamination (outside)	21	10.3
Contamination (building fabric)	7	3.4
Inadequate ventilation	98	48.3
Hypersensitivity pneumonitis	6	3.0
Cigarette smoking	4	2.0
Humidity	9	4.4
Noise/illumination	2	1.0
Scabies	1	0.5
Unknown	19	9.4
Total:	203	100.0

The malfunction of airflow mechanisms and the stratification of air, a condition in which a large part of the fresh air travels along the ceiling and fails to mix fully with air at the breathing zone, are among the other causes of inadequate IAQ (14). Various contaminant sources from inside or outside were the causes of the IAQ problems in 30% of the investigations conducted by Hughes and O'Brien. Many of these problems related to the poor design of the ventilation systems (such as air intake and infiltration from garages or congested traffic areas). In 10% of the investigations, the problems were attributed to annoying noise, illumination, or humidity. In 20% of the investigations, no specific cause could be identified.

More recent investigations of 150 buildings by NIOSH resulted in similar findings (15,16). These findings agree with the review of building-related illnesses by the Health and Welfare Canada (HWC), an organization engaged in a health hazard evaluation program that now includes IAQ. A recent review of 94 building investigations found problems with the ventilation system in 64 cases, reentry of building exhaust or the entry of motor vehicle exhaust in nine cases, and emissions from glues and adhesives in two cases (refer to Table 5-5). Although all of these investigations evaluated the possible role of ETS, the excessive use of photocopy machines or the presence of tobacco smoke were suspected of being the source of complaints in only five cases.

The results from investigations of health and comfort complaints in sick buildings indicate a wide range of contributing causes to inadequate IAQ. The findings from approximately 450 building investigations have failed to establish a significant relationship between ETS and building-related health and comfort problems. With the exception of hazardous building materials such as asbestos, inadequate ventilation is by far the most dominant problem facing the building owners and operators. The most severe secondary cause appears to be the contamination of fresh air systems by outside vehicle and restaurant exhaust or the reentry of building exhaust.

Although ETS is a minor contributor to the IAQ problems, it is often the one that is most blamed. ETS is the substance that can be visually seen and has a characteristic odor. If people are exposed to high concentrations of ETS, they may also be exposed to high concentrations of other air contaminants, some of which have been proven to cause serious health problems. The presence of visible ETS indoors is an indicator of a ventilation problem that should be studied and corrected immediately.

Table 5-5. Suspected Causes of Indoor Air Quality Problems (Health and Welfare Canada, Medical Services Branch (17)).

Suspected Cause	No. of Investigations	Percent of Total
Inadequate Ventilation	64	68
- Poor air circulation		
- Inadequate outdoor air (CO_2 > 800 ppm)		
- Poor temperature/humidity control		
Outdoor Contaminants	9	10
- Reentry of building exhaust		
- Motor vehicle exhaust		
Indoor Contaminants	5	5
- Photocopy machines		
- Tobacco smoke		
Building Fabric	2	2
- Glues and adhesives		
- Formaldehyde and organic		
Biological Contaminants	0	0
No Problem Identified	14	15
Total:	94	100.0

References

1. Sterling, D.A. "Volatile Organic Compounds in Indoor Air: An Overview of Sources, Concentrations, and Health Effects." *Indoor Air and Human Health* by Gammage, R.B. and Kaye, S.V., 1985.

2. Tichenor, B.A. and Mason, M.A. "Organic Emissions from Consumer Products and Building Materials to the Indoor Environment." *JAPCA*, 38. pp. 264 - 269, 1988.

3. Tancrede, M.; Wilson, R.; Zeise, L.; and Crouch, E.A. "The Carcinogenic Risk of Some Organic Vapors Indoors: A Theoretical Survey." *Atmospheric Environ.*, 21(10), pp. 2187 - 2205, 1987.

4. Sterling, T.D.; Collett, C.W.; and Sterling, E.M. "Environmental Tobacco Smoke and Indoor Air Quality in Modern Office Work Environments." *Journal of Occ. Medicine*, Vol. 29, No. 1, 1987.

5. National Academy of Sciences/National Research Council. Indoor Pollutants, National Academy Press, Washington, DC, 1981.

6. Sterling, T.D.; Dimich, H.; and Kobayashi, D. "Indoor By-product Levels of Tobacco Smoke: A Critical Review of the Literature." *JAPCA*, 32, pp. 250-259, 1982.

7. Turiel, I. Indoor Air Quality and Health, Stanford, CA, Stanford University Press, 1985.

8. National Academy of Science. "Environmental Tobacco Smoke, Measuring Exposures and Assessing Health Effects." National Academy Press, Washington, DC, 1986.

9. Arundel, A.; Sterling, T.; and Weinkam, J. "Exposure and Risk-Based Estimates of Never Smoking Lung Cancer Deaths in the U.S. in 1980 from Exposure to ETS." Proceedings

of the Indoor and Ambient Air Quality Conference, London, pp. 242-251, 1988.

10. Sterling, T.D. and Mueller, B. "Concentrations of Nicotine, RSP, CO and CO_2 in Non-Smoking Areas of Offices Ventilated by Air Recirculated from Smoking Designated Areas." *Am. Ind. Hyg. Assoc. J,* 49(9), pp. 423 - 426., 1988.

11. Balter, N.J.; Eatough, D.J.; and Schwartz, S.L. "Application of Physiological Pharmacokinetic Modeling to the Design of Human Exposure Studies with Environmental Tobacco Smoke." Proceedings of the Indoor and Ambient Air Quality Conference, London, pp. 179 - 188, 1988.

12. U.S. Surgeon General. "The Health Consequences of Involuntary Smoking." U.S. Department of Health and Human Services, Rockville, MD, 1986.

13. Melius, J.; Wallingford K.; Keenylyside, R.; and Carpenter, J. "Indoor Air Quality - The NIOSH Experience." *Ann. Am. Conf. Gov. Ind. Hyg.*, 10, pp. 3-7, 1984.

14. Hughes, R.R. and O'Brien, D.M. "Evaluation of Building Ventilation Systems." *American Indust. Hygiene Asso. Journal*, 47(4), pp. 207-213, 1986.

15. Carpenter, J. Hazard Evaluation and Technical Assistance Branch, National Institute for Occupational Safety and Health, Cincinnati, OH, 1986.

16. Wallingford, K.M. and Carpenter, J. "Field Experience Overview: Investigation Sources of Indoor Air Quality Problems in Office Buildings." Proceedings IAQ '86: Managing Indoor Air for Health and Energy Conservation, Atlanta, GA, pp. 448-458, 1986.

17. Kirkbride, J. "Sick Building Syndrome: Causes and Effects." Health and Welfare Canada, Ottawa, Ontario, Canada, 1985.

18. Walkinshaw, D. "Indoor Air Quality in Cold Climates, Hazard and Abatement Measures." *JAPCA*, 36, pp. 235-241, 1986.

6. SICK BUILDINGS: PHYSICAL AND PSYCHOLOGICAL EFFECTS ON HUMAN HEALTH AND PREVENTIVE MEASURES

Elia M. Sterling
Director of Building Research, Theodor D. Sterling and Associates, Ltd.
Vancouver, B.C., Canada

Page

6.1 Introduction

The term "sick building" is often associated with buildings in which a majority of occupants experience a variety of health and comfort problems for which no specific cause can be identified (1). Other cases of indoor air quality (IAQ) problems may be related to building-related illness (BRI) in which a known agent or pollutant is involved. Health complaints from occupants often include irritation of the eyes, nose, throat, upper respiratory system, headache, and general fatigue. This complex set of symptoms experienced by the occupants of modern buildings has been named as the sick-building syndrome (SBS) or the tight-building syndrome (TBS) (2), causing substantial increase in absenteeism and, therefore, loss of productivity among occupants suffering from SBS (3, 4). The majority of these sick buildings, constructed in the past ten years, are well-sealed, mechanically ventilated and air-conditioned and have few windows that can be opened. Several reviews (5, 6) have documented that sealed, air-conditioned buildings such as modern office buildings, contain a wide variety of pollutants often at very high concentrations (7).

To date, a large number of sick buildings has been investigated by the government agencies and independent researchers. Although most studies on sick buildings have been inconclusive, there exists a substantial amount of data in both the published and unpublished forms. These data include such parameters as IAQ, ventilation, lighting, acoustics, and reported effects on the health and comfort of occupants as well as research and instrumentation. A careful review of these data in addition to experience gained from numerous other investigations can be quite helpful in developing a systematic approach on how to diagnose a sick building, identify the cause of problems, and prescribe a course of action designed to correct the situation.

Case studies and other research have identified the following nine features common to unhealthy buildings (8):

- **A sealed building envelope.** Generally, the amount of fresh air introduced into a sealed, mechanically controlled building is minimized because it is energy efficient to recirculate as much of the building air as possible.

- **Heating, ventilating and air-conditioning.** The mechanical system helps the distribution of many indoor air pollutants generated by materials and equipment in a building. It may also incubate and spread fungi, bacteria, and viruses.

- **Location of vents and exhausts.** Air supply vents can introduce outdoor air contaminants into a building. For example, supply vents located near a busy street, parking garages, or freeways are often the source of entry for motor vehicle exhaust. Also, inadequate placement of supply and exhaust vents can prevent exhaust from escaping.

- **Location of ventilation diffusers.** Both the inlet and exhaust diffusers are commonly located in the ceiling, which often creates stratification and shortcircuiting of supply air at the ceiling resulting in dead air and poor circulation.

- **Lack of individual control over environmental conditions.** Not everyone is equally comfortable in the same indoor environment. Elimination of the possibility to change the environment may contribute to discomfort, stress, and other minor health problems.

- **Use of new materials and equipment.** Synthetic materials, modern office equipment, industrial soaps, detergents and waxes used for maintenance generate many irritating and sometimes toxic fumes and dusts including formaldehyde, hydrocarbons, amines, ozone and respirable particulates.

- **Fluorescent lamps.** The fluorescent lamps emit ultraviolet light and may provide energy for photochemical reactions among pollutants, thus creating indoor smog.

- **Parking garages, restaurants and other non-office space use.** Many parking garages, access to transportation such as buses and subways, restaurants, health clubs, laundry and recreation facilities may add substantial amounts of combustion byproducts.

- **Energy conservation methods.** Most energy conservation methods usually involve reduction of fresh air ventilation rates, which increases the rate of accumulation of pollutants by reducing the volume of air exhausted. The efficiency of standard air filters is reduced substantially as the ventilation air velocity is lowered. Many buildings use a variable-air-volume (VAV) system,

> which introduces fresh air only when cooling or heating is required. Consequently, occupants of a building, often complain of stale, stuffy air which indicates insufficient ventilation.

High cost of fuels in the 1970's placed immediate pressure on energy conservation. Building construction, maintenance, and service practices and standards were revised to allow energy conservation. The majority of new office buildings today are being designed and built to comply with the revised environmental standards to achieve energy conservation goals. Also, many existing contemporary office buildings are being retrofitted to reduce the amount of energy consumed. The cost in terms of human health, comfort, and productivity that may result from the revised environmental standards in office buildings are still unclear.

6.2 Health Effects

The common symptoms of SBS include headache, eye problems, nasal problems, throat problems, fatigue and lethargy, chest problems, skin problems, and problems in maintaining concentration among the occupants of a building.

Research conducted to date has not isolated a causal agent or agents for SBS. Studies comparing ambient conditions in air-conditioned and naturally ventilated buildings have typically found little difference in any of the environmental parameters measured (9,10). One of the major difficulties associated with such comparisons is the lack of any clear consensus on both defining symptoms of the SBS and the rate of occurrence at which such symptoms become indicative of an SBS.

In a survey administered to the air-conditioned and naturally ventilated buildings, the symptoms included those of frequently encountered in SBS as shown in Table 6-1 (11). However, it may be unwise to generalize the results of this survey to all types of buildings. More extensive research in a larger number of buildings with a wide range of ventilation systems is essential. Investigation of such symptoms in different types of office buildings may be useful in the development of a standard diagnostic questionnaire for identification of sick buildings.

Table 6-1. Comparison of Health Complaints in Air-Conditioned and Naturally Ventilated Office Buildings.

Symptom	Air-Conditioned Offices (% usually)[A]	Naturally Ventilated Offices (% usually)[A]
Sleepiness	69.2	44.5
Fatigue	68.0	52.4
Headache	67.2	50.5
Eye irritation	52.1	45.9
Concentration problems	50.9	41.2
Cold/flu symptoms	50.2	32.4
Sore throat	47.9	28.3
Nose irritation	45.5	26.5
Focusing problems	42.9	28.8
Backache	41.8	41.4
Neckache	41.2	39.5
Cold extremities	40.7	38.8
Tension	36.1	33.1
Skin dryness	29.9	16.7
Depression	25.1	25.2
Dizziness	23.6	15.5
Muscular aches	21.1	17.2
Weakness	20.3	9.1
Nausea	19.4	7.8
Respiratory problems	12.2	5.7
Chest tightness	9.8	6.8
Fever	8.1	2.0

A: % usually = % sometimes + % always

According to several investigations, certain syndromes with recognizable symptoms may occur in offices, homes and hospitals in response to specific toxic dusts, fumes, or viable microorganisms. Some of these factors are identified to be (12): (a) dry detergent residues, (b) fibrous glass dust from ductwork, (c) formaldehyde off-gassing from insulation, (d) photochemical smog formation, and (e) diseases from viable microorganisms located in duct systems, cooling towers, or humidification chambers. Cigarette smoking is sometimes associated with such symptoms.

6.3 Preventive Measures

In more than 350 investigations conducted by the National Institute of Occupational Safety and Health (NIOSH) in the United States (13), cigarette smoking was suggested as a suspected cause in only 2% of the investigations. By far the most prevalent problem was that of inadequate ventilation (48.3%). The most common cause of inadequate ventilation is the diminished intake of fresh air into the air circulation system, usually to conserve energy and save on cost of building operations. A recent review of 94 building investigations by Health and Welfare Canada (HWC) (14) also found problems with ventilation systems in high percentage (68%). In this study, the combined category of photocopy machines and tobacco smoke was associated with SBS in only 5% of all cases.

The detailed monitoring of ventilation and indoor air contaminants necessary to document possible mechanical system inadequacies can be time-consuming and expensive. However, a multiphase program of gathering information from building occupants and maintenance personnel, combined with the measurements of specific indoor air pollutants and the inspection of easily observable ventilation parameters, can provide a timely and cost-effective method of investigation. Such an approach may provide a practical means of making judgments about the adequacy of performance of a building ventilation system and other potential sources of indoor air contaminants. The following approach for building performance evaluation consists of five phases, and is used to (15): (a) determine whether an IAQ problem exists in a building, (b) identify the probable causes of the IAQ problem, (c) design and implement modifications to alleviate the problem, and (d) reevaluate IAQ conditions after modifications have been made to test the effectiveness of the design solutions.

Phase 1:

Phase 1 consists of a checklist that should be completed by the building owner, operator, or representative such as the maintenance personnel. This checklist contains information on the architectural and mechanical system design and performance (including maintenance), use of the building (both by employees and visitors), workspace design and layout, equipment use and occupant health, and the comfort concerns and complaints. Review of the information from the building owner/operator provides basic information about the mechanical system performance and occupant problems.

Phase 2:

Phase 2 includes two parts:

a. Administration of an "Office Work Environment Survey" questionnaire to all building occupants as a method for documenting IAQ problems and health symptoms experienced by the occupants and to locate areas where complaints are more acute for detailed monitoring in Phases 3 and 4, if required.

b. A walkthrough evaluation of the building that would be conducted by the state government officers. It includes an inspection of the mechanical systems and review of the mechanical and architectural plans to obtain an overview of building performance. The review of plans also allows comparison of the design specifications of the mechanical systems with the established standards such as ASHRAE Standard 55-1981 and ASHRAE Standard 62-1989.

Phase 3:

Phase 3 consists of measurements of the selected major IAQ parameters, with locations for air-sampling determined by the results from Phase 2. Some of these parameters are:

- Carbon dioxide as an indicator of the buildup of contaminants generated indoors;

- Carbon monoxide as an indicator of combustion byproducts infiltrating the building, especially from parking garages or other sources of indoor combustion;

- Temperature and relative humidity as indicators of occupant thermal comfort conditions; and

- Respirable particulates as an indicator of ventilation filtration system efficiency and complaints related to environmental tobacco smoke (ETS).

Each of the parameters can be measured using a portable direct reading instrument, which can be easily carried in a building. Other parameters such as formaldehyde, ozone, and microorganisms may also be used in Phase 3 as determined by the data obtained from Phases 1 and 2. Present measurement methods for these additional parameters often involve an extensive laboratory analysis of collected samples. However, measuring microorganisms, a simple method for providing counts of total microbes is under development and when available, following field testing, may be incorporated into Phase 3. Additional information on sampling equipment and screening techniques may be found in Chapter 15, Part 3.

Phase 4:

Phase 4 is a detailed ventilation evaluation of a building using smoke pencil tests, tracer gas, and air flow measurements to determine:

- Total building air exchange rates;

- Floor air exchange rates; and

- Patterns of air movement throughout the building including air leakage from potential indoor air pollutant sources, such as parking garages.

Following Phase 4, investigators will be in a position to make recommendations based on the combined results of Phases 1 through 4.

Phase 5:

Phase 5 represents the implementation of design solutions in a building. Phase 5 may be reached at the completion of any of the first four phases, as dictated by the results of specific evaluations.

After modifications have been made, a vital further step in total building performance investigation is a reevaluation of IAQ conditions to determine whether recommended modifications have been effective. Reevaluation begins with the readministration of Phase 1. If modifications to the building have been effective, the completion of Phase 1 will indicate that no further IAQ problems exist. If IAQ problems still exist in the building, a strategy is to continue through the components phases until IAQ and related ventilation system performance problems are eliminated.

The five-phase approach outlined above has been developed as a practical strategy to locate and identify the probable causes of IAQ problems in a building in a timely and cost-effective manner, and to formulate retrofit actions to improve conditions.

The adoption of a standard approach to investigate buildings may be beneficial to both the researchers and building owners/ occupants because the findings from different buildings may be directly compared. In addition, as more investigations are undertaken, the baseline data from health and comfort complaints, and measured IAQ and ventilation parameters would be developed with which future investigations may be compared.

References

1. Sterling, E.M.; McIntyre, E.D.; Collett, C.W.; Sterling, T.D.; and Meredith, J. "Sick Buildings: Case Studies of Tight Building Syndrome and IAQ Investigations in Modern Office Buildings." *Environ. Health Review*, 1985.

2. Hicks, J.B. "Tight Building Syndrome." *Occ. Health and Safety Magazine*, pp. 1-12, Jan., 1984.

3. Sterling, E.M. and Sterling, T.D. "The Impact of Different Ventilation Levels and Fluorescent Lighting Types on Building Illness." *Canadian J. of Public Health*, 74, pp. 385-392, 1983.

4. Baron, L.I. and Sterling, E.M. "Does Indoor Air Pollution Affect Office Productivity?" Proceedings: 3rd International Symposium on Building Economics, National Research Council of Canada, NRCC 23309, 5, pp. 22-28, 1983.

5. Sterling, T.D. and Kobayashi, D.M. "Exposure to Pollutants in Enclosed Living Spaces." *Env. Res.*, 13, pp. 1-35, 1977.

6. Yocum, J.E.; Clink, W.L.; and Cote, W.A. "Indoor/Outdoor Air Quality Relationships." *JAPCA*, 25, pp. 251-259, 1971.

7. Sterling, E.M. "Indoor Air Quality - Total Environment Performance: Comfort and Productivity Issues in Modern Office Buildings." *Resource*, Feb., 1986.

8. Sterling, E.M. and McIntyre, E.D. "Summary of Characteristics Common to Sick Buildings." *Environ. Health Review*, 1985.

9. Robertson, A.S.; Burge, P.S.; Hedge, A.; Sims, J.; Gill, F.S.; Finnegan, M.J.; Pickering, C.A.; and Dalton, G. "Comparison of Health Problems Related to Work and Environmental Measurements in Two Office Buildings with Different Ventilation Systems." *British Medical Journal*, 291, pp. 373-376, 1985.

10. Turiel, I.; Hollowell, C.D.; Miksch, R.R.; Rudy, J.V.; and Young, R.A. "The Effects of Reduced Ventilation on Indoor Air Quality in an Office Building." *Atmospheric Environ.*, 17 (1), pp. 51-64, 1983.

11. Sterling, E.M.; Hedge, A.; and Sterling, T.D. "Building Illness Indices Based on Questionnaire Responses." Proceedings of Indoor Air Quality '86: Managing Indoor Air for Health and Energy Conservation, 1986.

12. Sterling, T.D.; Sterling, E.M.; and Dimich-Ward, H. "Building Illness in the White-Collar Workplace." *International Journal of Health Services*, Vol. 13, No. 2, 1983.

13. Melius, J.; Wallingford, K.; Keenleyside, R.; and Carpenter, J. "Indoor Air Quality: The NIOSH Experience." Annual Meeting of the American Conference of Government Industrial Hygienists, Atlanta, GA, 1984.

14. Kirkbride, J. "Sick Building Syndrome: Causes and Effects." Health and Welfare Canada, Ottawa, 1985.

15. Sterling, E.M.; Collett, C.W.; and Meredith, J. "A Five Phase Strategy for Diagnosing Air Quality and Related Ventilation Problems in Commercial/Large Buildings." *APCA*, NY, 1987.

PART 2

ENGINEERING SOLUTIONS TO INDOOR AIR QUALITY PROBLEMS

7. CONTROL OF INDOOR AIR POLLUTANTS

Milton Meckler, P.E.
President, The Meckler Group
Encino, California

and

Elia M. Sterling
Director of Building Research, Theodor D. Sterling
and Associates, Ltd.
Vancouver, B.C., Canada

7.1 Introduction

In this chapter, we will take a generalized approach to controlling indoor air contaminants. The methods that will be explored are: (a) source removal, (b) ventilation, and (c) air-cleaning. Also at the latter part of the chapter, we will discuss a method for variable-air-volume (VAV) systems which involves additional filtration and recirculation of ventilation air to provide enhanced overall filtration.

7.2 Source Removal

Although ASHRAE Standard 62-1989 relies mainly on a ventilation (dilution) control strategy, it refers to direct source control and source removal by chemical or physical means such as exhaust or recirculation employing effective air-cleaning methods.

Significant sources of outdoor air pollutants such as motor vehicle exhaust from streets and parking lots, loading ducts, emissions from cooling towers, roofing operation emissions, and exhausts of nearby buildings or industrial sources should be avoided. This is a special problem in buildings where stack effect draws contaminants from these areas into the occupied spaces. Where soils contain significant concentrations of radon, ventilation practices which place crawl spaces, basements or underground ductwork below atmospheric pressure will tend to increase radon concentrations in buildings and should be also avoided.

Use of dilution ventilation for controlling airborne health hazards for large quantities of contaminants, elevated local concentrations, or highly toxic contaminants, should not be employed. These situations would require extremely large volumes of dilution air with questionable effectiveness. Instead, properly designed exhausts, forced exhaust air near a major source of toxic contaminants, will be a more efficient and cost-effective solution. Ventilating duct and plenums should also be constructed and maintained to minimize the growth and dissemination of microorganisms through the ventilation system. Lined ducts should be avoided especially where condensation may occur.

Selection of gaseous contaminant control equipment for recirculation systems should consider the concentration, toxicity, annoyance, odor properties of the contaminants present, and the levels to which these must be reduced to be effective in maintaining proper indoor air quality (IAQ). Equally, the performance of these equipment often depends strongly on the physical and chemical properties of each contaminant present, temperature and humidity of the air, air velocity through the equipment, and the loading capacity.

Contamination of outside air intakes may be caused by standing water; nests or feathers near outside air intakes; and polluted outside air from adjacent buildings, parking garages, streets or freeways. Among the other contamination sources to avoid are unfiltered outside air intakes or dirty outside air filters.

7.3 Effect of Ventilation on Indoor Air Quality

Until recently, the most widely preferred strategy for reducing indoor air contamination was to dilute with fresh (make-up) air from outside and use of appropriate filtration. However, increased energy costs for heating and cooling make-up air have made this option less attractive (1). Reductions in both the ventilation and energy use have been accompanied by the increasing health and comfort complaints among occupants, often known as the sick-building syndrome (SBS).

Ventilation system design is based on the projected amount of air needed to provide acceptable IAQ and comfortable conditions to building occupants (2). These projected values are based on building design factors and best estimates of heating, cooling, and occupancy loads. The total amount of air, amount of outdoor air, and the amount of recirculated air are specified to satisfy the projected requirements of a building based on the estimated number of occupants. However, after the completion and during lifetime of a building, actual loads may exceed or fall short of these parameters. Ventilation parameters are usually expressed as overall measures that best represent the actual loads. The most commonly used parameter for actual load is human occupancy. Using the actual number of occupants as a base, a set of parameters measuring ventilation capacity in terms of the total, outdoor, and the recirculated amount of air supplied to each person occupying the building may be obtained. Thus, there are two sets of ventilation parameters: the designed-in parameters based on estimates, and parameters calculated from actual use.

For purposes of ventilation to control IAQ, it is of great importance to determine: (a) if there are discrepancies between different ventilation measures, especially among the designed-in parameters and those affected by actual occupancy; and (b) if there are such discrepancies and, which are the most suitable or effective for controlling IAQ (2).

During the last few years, several buildings have been studied to determine the relationship of indoor air pollutants to either health and comfort complaints of occupants or to building characteristics. These studies contain vast amounts of data about the features, contents, and the possible use of the buildings; characteristics and reactions of occupants; ventilation; and the ambient environments.

Ventilation is expressed in various units (3). The most commonly used units are air changes per hour (ach) and cubic feet per minute (cfm) or liters per second (L/s). Ach is the measure of total air exchange in an enclosed area based on time, and cfm (or L/s) is the volumetric flow rate of air. Requirements for ventilation of enclosures listed in ASHRAE Standard 62-1989 are calculated on the basis of occupant density (cfm/person or L/s/person), and in some cases based on other variables (e.g., floor space or number of beds in hospitals). The total amount of air supplied (cfm) is a mixture of recirculated air (cfm) and the fresh outside air (cfm). However, it is not always known what the occupant load will be during the life span of the building. Therefore, the occupant load is often estimated on a square foot basis (4). Consequently, a number of ventilation parameters are designed-in to satisfy the estimated ventilation requirements. The basic designed-in parameter is the total supply air (cfm). The amount of fresh make-up air (cfm) and recirculated air (cfm) are regulated by dampers, and their total is expressed as the total supply air (cfm).

The designed-in total supply air (cfm) varies with the size of the building, its basic characteristics, number of occupants, location, and other expected use and performance (3). In general, total supply air is larger with larger buildings that are expected to have greater use and occupancy. However, any standard-setting attempt needs to be aware of the degree to which these designed-in ventilation parameters correlate with each other and with parameters affected by the actual use and occupancy.

There are a number of ventilation parameters that can be defined by expressing the total, fresh, and recirculated air with respect to actual occupancy and activities in a building. Perhaps the most important are the ones related to actual occupancy. These can be expressed as an operational total supply air (cfm/person), fresh make-up air (cfm/person), and recirculated air (cfm/person). The total supply air (cfm) is equal to total supply air (cfm) divided by the actual number of occupants, and fresh make-up air (cfm/person) and recirculated air (cfm/person) are calculated likewise.

ASHRAE Standard 62-1981 and ASHRAE Standard 62-73 are based primarily on the chamber studies such as those conducted by Yaglou in 1936 (5) or more recently by Cain in 1983 (6). The results were reported in fresh air ventilation (cfm/person) needed to provide an odor-free environment (7). In the past 50 years, significant advances have been made in both the IAQ and ventilation measurement technology. Also in the past ten years, both the ventilation and IAQ have been measured simultaneously in investigative studies of many modern, sealed-buildings. For controlling and setting ventilation standards to maintain proper IAQ, it is of great importance to determine: (a) the strength of relationship between the ventilation and contaminant concentrations, and (b) the most effective measure of ventilation for controlling IAQ.

7.4 Air-Cleaning Systems

Some indoor air contaminants (principally particulates) can be removed effectively by commercially available filtration and removal technology. Several design considerations affect the performance of an air-cleaning system such as the type of air cleaner used, its location in the system, and the amount of air passed through the air cleaner (8).

Particles are the easiest contaminants to remove from the air. Non-specific particle air cleaners are: (a) media filters, (b) electrostatic air cleaners, (c) adsorbers, (d) centrifugal separators, and (e) air washers and the other absorbers. The following discussion of the different types of air-cleaning systems is provided to give the reader some idea of the options available.

7.4.1 Media Filters

Media filters are commonly used in building heating, ventilating, and airconditioning (HVAC) systems (8). They consist of a porous material, frequently fibrous glass through which the air is forced. The small air passages cause particles to be strained out by impact on the filter media. These filters range in performance from the dust-stop filters used in most warm-air furnaces to remove lint and large dust particles, to high efficiency particulate air (HEPA) filters that can remove submicron particles. Particulate matter collected on a media filter alters the performance of the filter. Filtration effectiveness tends to increase with filter loading, but the increased pressure loss across the filter reduces the airflow. Thus, while filtration efficiency may increase with use, the amount of air filtered tends to decrease. Media filters require regular maintenance with cleaning or replacement.

7.4.2 Electrostatic Air Cleaners

Electrostatic air cleaners (EACs) offer many desirable features to the user. Particles passing through the charging section of an EAC receive an electric charge which causes them to be attracted to the oppositely charged plates in the collector section. EACs can have a very high efficiency for collecting particles in the range of 0.01 micron to 5 microns (8). EACs generally do not restrict airflows and have a very small pressure loss. Regular cleaning is required to remove the collected materials from the collector plates. This is accomplished by washing the collector section of the filter. The whole plate is removed and washed in residential and small commercial cleaners. Wash-in-place designs are available in larger EACs.

7.4.3 Adsorbers

Activated carbon and activated alumina are adsorbers (8). Activated carbon has a very large surface area due to fine voids in its molecular structure. Adsorption of gases occurs by adhesion of gas molecules to the surface of the carbon structure; thus activated carbon can remove certain gases and vapors. In general, the effectiveness is related to the size of the molecules. Activated carbon is quite effective in removing some odor molecules (e.g., hydrogen sulfide) but is ineffective in removing small molecules such as carbon monoxide. Activated alumina is also a porous material. Since alumina is considerably more expensive, it is not usually used in the pure form. However, activated alumina saturated with potassium permanganate can be used as a chemically active air cleaner. Potassium permanganate is a strong agent that is very effective in oxidizing molecules that cause odor. Activated alumina treated with potassium permanganate is quite effective in removing formaldehyde which is converted to carbon dioxide and water.

Activated carbon and activated alumina require regular replacement. Activated carbon can be regenerated with steam. However, this is not considered to be a viable process for air cleaners used in building HVAC systems. The potassium permanganate in activated alumina filters is a consumable reagent and, the filter must be replaced when the reagent is exhausted.

7.4.4 Centrifugal Separators

Centrifugal separators (cyclone) depend on the difference between the density of the air and density of the particles and the particle size. Very small particles have relatively large surface areas for their masses with increased aerodynamic drag in a moving airstream. Therefore, centrifugal separators are ineffective in removing very fine particles and used mainly for dust removal in industrial applications.

7.4.5 Air Washers and Other Absorbers

Air washers are also used as the washing fluid to remove gases such as sulfur dioxide from boiler gases in power plants. Air washer systems are used in atomic submarines to remove carbon dioxide. However, the high cost and need for regular maintenance make air washer systems unattractive for use in commercial buildings.

In a dehumidification system, 100% of the air to be dehumidified is exposed to a highly stable, non-toxic moisture absorbing liquid desiccant (e.g., lithium chloride). The amount of moisture which a liquid desiccant will take out of the air is directly related to the concentration and temperature of the solution. As the solution is cooled, the dehumidification system produces dry air. This unique operating principle of a dehumidification system offers far more than precise humidity control. Because of its chemical composition, the liquid desiccant is highly biocidal and automatically removes up to 99% of airborne microorganisms including bacteria. The use of liquid desiccants to enhance IAQ is explored fully in an integrated desiccant cold air distribution (IDCAD) system in Chapter 11, Part 2. Also, the role of solid desiccants in IAQ may be found in Chapter 12, Part 2.

7.5 Use of Air Cleaners to Provide Enhanced Overall Filtration

Lately, improved air filters and air-cleaning systems that remove particulate and gaseous contaminants from outside air and return air have received a great deal of interest. It may be instructive to examine the following considerations (8):

a. Under what conditions can placement of one or more air filters at the air-handling (terminal) unit or in adjacent ductwork affect the performance of overall air filtration?

b. How can the use of air recirculation at part-load enhance the overall performance of VAV air distribution systems to maintain proper IAQ at reduced air rates?

One of the methods, for VAV systems, involves the additional filtration and recirculation of ventilation air to provide enhanced overall filtration (9) to compensate for the reduced supply airflow. VAV systems should be controlled so that the outside air drawn into the system is in the same proportion to the total amount of air reduced. However, this may not be practical due to increased cost.

Overall net air filtration efficiency, E_f, of one or more air filters located in positions A and/or B as shown in Figure 7-1, can have a significant effect on the outdoor air rate, V_o. The IAQ performance of tested VAV systems is particularly sensitive to the selection of air cleaners, especially where increased air recirculation is contemplated as a design strategy for lowering outdoor air requirements (ventilation). Since VAV air distribution systems may be subject to IAQ problems, it may be instructive to examine how air recirculation and judicious placement of air cleaners can be used to improve IAQ in spaces served year-round.

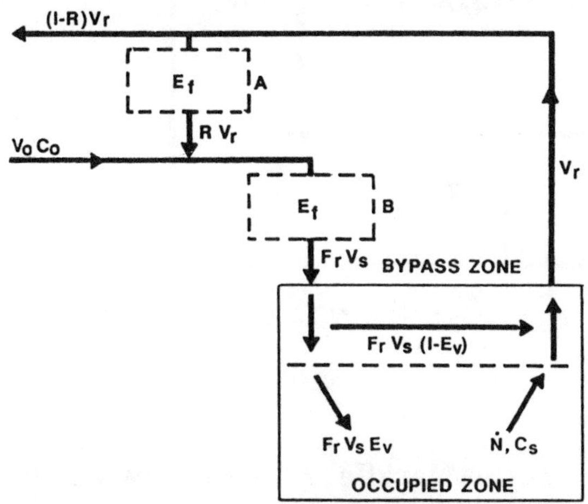

Figure 7-1. Alternate Air Cleaner Locations (ASHRAE Standard 62-1989).

Let us assume that we wish to serve the temperature control zone shown in Figure 7-2. Let us further assume that the efficiency of the proposed primary air filter is designated by E_1, and that the designer is considering some further contaminant control enhancement by adding a secondary air filter of efficiency E_2. This secondary filter can be placed either ahead of the primary air filter in position B as shown in Figure 7-1, or at another location. There are two proposed alternative air filtration scenarios:

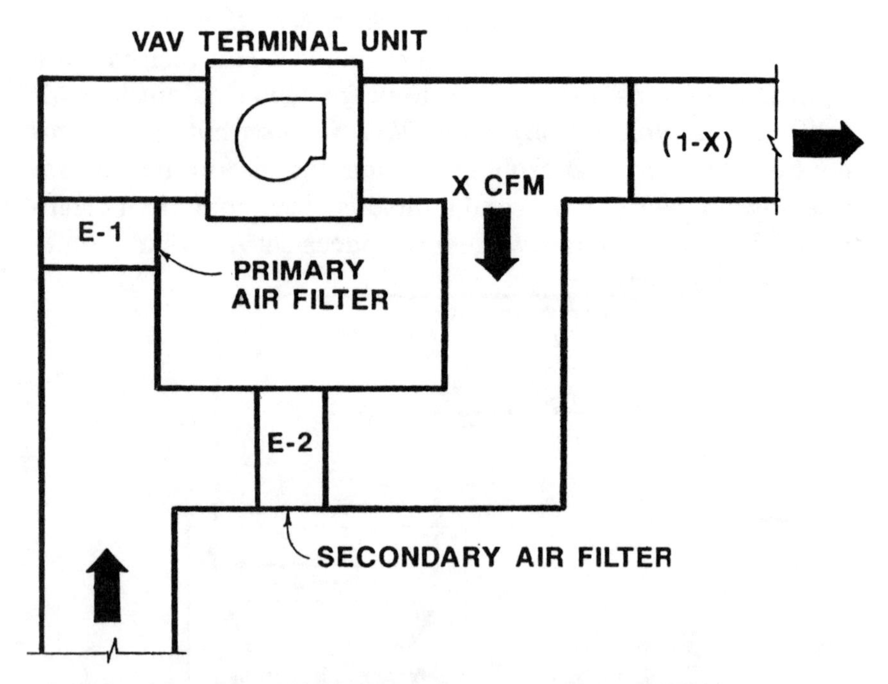

Figure 7-2. **Series/Parallel Bypass Flow Air Filtration (Scenario 1: Recycle Series/Parallel Filtration System).**

Scenario 1 contains a ceiling-mounted unitary water-cooled air conditioner (or heat pump) with a constant-speed supply fan equipped with a bypass damper control. This control feature is capable of sensing static pressure buildup in the main supply air duct, as shown in Figure 7-3. Individual zone control dampers in their respective spaces modulate closed-in response to net space demands for heating and cooling. These zone control dampers are commercially available.

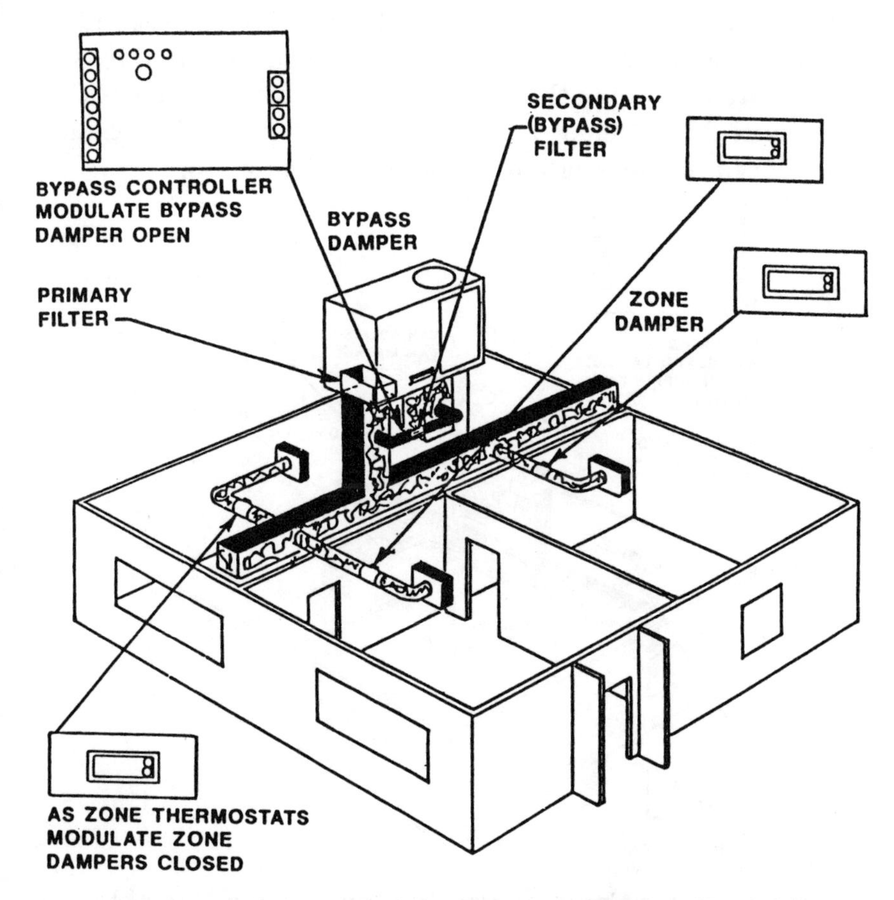

Figure 7-3. Air-Conditioning System With a Variable-Air-Volume Unit (Filters in Both Return and Bypass Positions, Scenario 1).

To achieve the overall desired VAV airflow rate to all zones served, supply air is automatically bypassed to the inlet, thereby providing a natural path for air recirculation where the volume control damper must add more or less air frictional resistance. Therefore, locating the secondary air filter in the bypass duct (provided that the primary air filter is located at position B) would improve overall air filtration at no additional energy cost.

Figure 7-2 shows locations of the primary air filter and the secondary air filter in a series/parallel bypass airflow or air recirculation configuration. Figure 7-4 illustrates an HVAC/space configuration similar to that shown in Figure 7-2 but differs in several important details. Note that the terminal VAV unit's air conditioner supply fan is of the variable-speed type responding directly to net space demands for VAV flow, also sensed by static pressure buildup in the main supply air duct, and identified as Scenario 2. In Scenario 2, a secondary air filter of efficiency E_2 is placed in the position as shown in Figure 7-5.

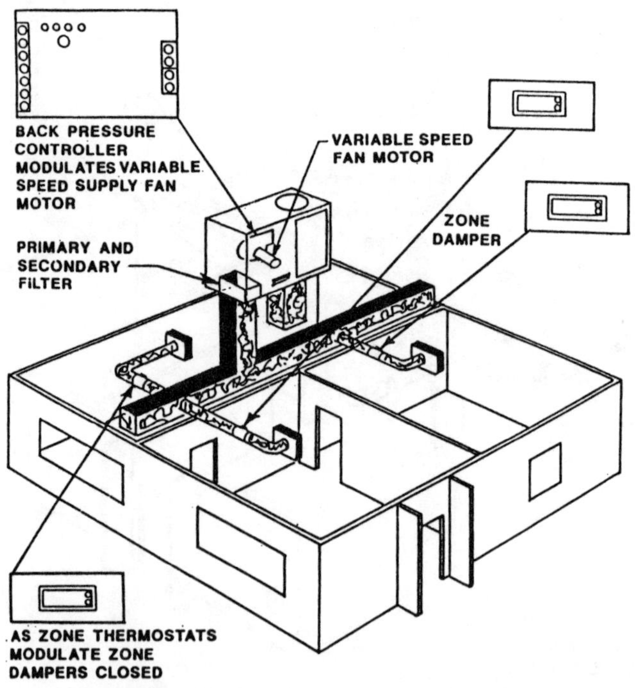

Figure 7-4. Air-Conditioning System With a Variable-Air-Volume Unit (Filter in Only Return Air Position, Scenario 2).

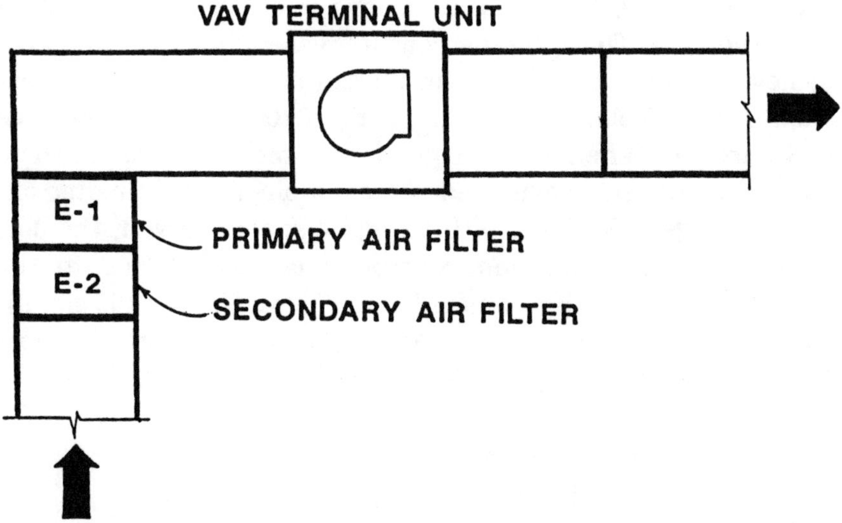

Figure 7-5. Series Flow Air Filtration System, Scenario 2.

In Scenario 1, when less than 100% (or design day cooling) supply air is needed, one can directly compute the fraction of the total supply air delivered by the terminal VAV unit fan that will automatically bypass through the secondary air filter. The quantity (1-X) represents the fraction of the total supply air delivered by the terminal VAV unit supply fan that is actually supplied to the conditioned spaces.

In Scenario 1, shown in Figure 7-2, one must first determine a seasonal average for X. If we define $E_{p/s}$ as the overall air filtration efficiency, we arrive at the following relationship:

$$E_{p/s} = 1 - \frac{(1 - E_1)}{1 + X\,[1 + (1 - E_1)(1 - E_2)]} \qquad (1)$$

Similarly, if we define E_S as the overall air filtration efficiency for Scenario 2, we arrive at a comparable relationship to that shown in Equation 1 and is:

$$E_S = 1 - (1 - E_1)(1 - E_2) \qquad (2)$$

Solving for either E_S or $E_{p/S}$ in Equations 1 and 2, one can directly substitute either value for the overall net air filtration efficiency, E_f. Then, the percent increase in air filtration efficiency of Scenario 1 to Scenario 2 can be determined. For example, one can plot $\{[(E_{p/S} - E_S) \div E_S] \times 100\}$ versus X as shown in Figure 7-6. Whenever X is greater than or equal to 0.1 some improvement can be expected. For example, at an estimated annual average value of X = 0.5 for the spaces served, the corresponding contamination control enhancement due to the proposed recirculation ranges from 35% to 65% and is significant. Note that the proposed bypass position, the pressure loss through the secondary air filter shown in Figure 7-2 results in no net energy penalty.

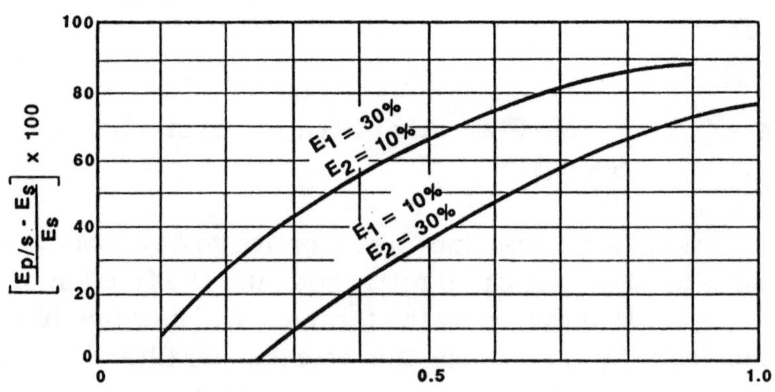

RECIRCULATION (BY-PASS) AIR FRACTION X

E_1 = FILTER EFFICIENCY OF SECONDARY FILTER LOCATED IN SCENARIO 2 BY-PASS DUCT POSITION

E_2 = FILTER EFFICIENCY OF PRIMARY AIR FILTER LOCATED IN MIXED OSA/RA POSITION

NOTE : FILTER EFFICIENCY DEFINED BY PERCENT CORRESPONDING TO ASHRAE STANDARD 52-76

Figure 7-6. Increase in Overall Filtration Efficiency vs. Recirculation Air Fraction X.

Air-cleaning systems that effectively remove the major indoor air contaminants can reduce the amount of outdoor air required. The use of recirculation combined with adequate filtration as a viable cost effective alternative to increasing outdoor air rates provides the HVAC designer with the means to substantiate his decisions when dealing with client concerns about increased energy consumption.

References

1. American Society of Heating, Refrigerating, and Air-Conditioning Engineers. "ASHRAE Position Statement on Indoor Air Quality." Atlanta, GA, 1982.

2. Sterling, E.M. and Sterling, T.D. "The Impact of Building Ventilation on Indoor Gaseous and Particulate Pollution in Office and Institutional Buildings." *Ventilation '85*, Elsevier Science Publishers B.V., Amsterdam, 1986.

3. Sterling, E.M. and Sterling, T.D. "Effects of Different Ventilation Parameters on Indoor Pollutants." Proceedings of an Engineering Foundation Conference on Management of Atmospheres in Tightly Enclosed Spaces., pp. 133-140, 1983.

4. Leaney, D. D.W. Thompson Ltd., Consulting Engineers, 1983.

5. Yaglou, C.P.; Riley, E.C.; and Coggins, D.I. "Ventilation Requirements." *ASHRAE Transactions*, 42, pp. 133-136, 1936.

6. Cain, W.S.; Leaderer, B.P.; Isseroff, R.; Berglund, L.G.; Huey, R.J.; Lipsitt, E.D.; and Perlman, D. "Ventilation Requirements in Buildings - I. Control of Occupancy Odor and Tobacco Smoke Odor." *Atmos. Env.*, 17(6), pp. 1183-1197, 1983.

7. Woods, J.E. "Ventilation, Health and Energy Consumption: A Status Report." *ASHRAE Journal*, pp. 33-39, July, 1979.

8. Meckler, M. and Janssen, J.E. "Use of Air Cleaners to Reduce Outdoor Air Requirements." Proceedings of the ASHRAE Conference, IAQ '88: Engineering Solutions to Indoor Air Problems." Atlanta, GA, pp. 130-147, April, 1988.

9. National Institute of Building Sciences. "Building Air Filtration and Ventilation." Proceedings: Program Planning Workshop, sponsored by Building Thermal Envelope Coordinating Council, 1985.

8. EVALUATION OF METHODS FOR MEASURING MAJOR INDOOR AIR POLLUTANTS

Frank Vaculik
Sr. Operation & Maintenance Engineer,
Public Works Canada
Ottawa, Ontario, Canada

Page

8.1 Introduction

Today, the building industry is faced with a major challenge in responding to tenant concerns about indoor air quality (IAQ). Their response is hampered by inadequate methods whereby IAQ may be measured and controlled on a continuous basis in compliance with the scientifically derived standards defining environmental air quality requirements. The purpose of this chapter is to introduce a method that may be used to overcome such difficulties in evaluating the performance of ventilation systems in buildings.

This chapter is directed to building operators who represent a group of individuals consisting of owners, property managers, physical plant engineers, and the operation and maintenance (O&M) personnel. The building operators must respond to the challenge of meeting IAQ requirements during the life span of their buildings. Their success, in turn, depends on the designers, contractors, and equipment suppliers who must furnish them with systems capable of complying with the current practical ventilation and air quality standards. Currently available air quality standards, such as ASHRAE Standard 62-1989, are based primarily on the subjective judgment of experienced professionals. These comfort-oriented standards are generally much more stringent than those required for personal safety and to protect the health of occupants. Current health and safety standards, published by the American Conference of Governmental and Industrial Hygienists (ACGIH), are primarily aimed at industrial applications.

Despite the inadequacy of knowledge and equipment, the building operators must respond to IAQ issues in accordance with the currently available standards. This applies to both the new and existing buildings regardless of the adequacy of their systems. The building operators must have the necessary means for correct assessment of performance. The carbon dioxide method of general ventilation control, that uses carbon dioxide (CO_2) as a surrogate, provides this means. It takes advantage of the fact that CO_2 is produced by occupants free of charge and, is a measure of a person's metabolism. The method is suitable to control ventilation in a ventilation zone consisting of several workstations with different ventilation requirements. It is applicable to both the new and existing buildings and ensures compliance with the ventilation standards. Although this method needs further refinement, it already offers building operators the necessary tools to adequately control IAQ in a more energy-efficient manner. This chapter explains the principles involved; discusses the effect of intervening ventilation variables; provides ventilation control strategies for several building systems; and offers suggestions for implementing, commissioning, and further research.

In most office buildings (institutional and commercial), the majority of indoor air contaminants are more likely to be generated throughout the space rather than in isolated areas. In these buildings, IAQ is controlled through a process of dilution of indoor air contaminants by outdoor air. This process is called general ventilation as opposed to local ventilation employed to control contaminants generated in relatively isolated areas. ASHRAE Standard 62-1989 prescribes rates of relatively clean outdoor air that are expected to keep IAQ at a certain comfort level acceptable to the majority of occupants.

The specific outdoor air rates are expressed in liters per second per person (L/s/person). This is a basic parameter that the building operators must monitor on a real-time basis for correct assessment of compliance with the general ventilation requirements. However, the expression of outdoor air rates is inadequate in dealing with the majority of occupants. An expression of air changes per hour is more readily understood for this purpose. The building operators administer the IAQ control clauses of the occupancy agreements more effectively with both measures available. This is essential because these agreements for owned or leased spaces are the fundamental business transactions of the real estate industry that the system designers and construction personnel depend on.

8.2 General Ventilation and Mixing

The general ventilation is a dilution process based on mixing of two fluids. Although mixing of fluids is a common process, it is not fully understood in building ventilation applications. The following discussion will explore the fundamentals of dilution or mixing process that is extremely important to the control of IAQ.

The simplest application of mixing is the one that is independent of time. When a small container is filled with fluids A and B and stirred, the concentration of each fluid in the mixture can be easily determined by using conservation of mass as A/M or

B/M, where M = A + B. For example, a 3-liter container of 2% milk contains 0.06 liter of butter fat or a 1-liter bottle of 40% rum contains 0.4 liter of pure alcohol.

The steady state mixing (continuous mixing) shown in Figure 8-1 is slightly more complex. In this, fluids A and B are mixed to produce fluid M, in which the concentration of either fluid A or B is determined as above. This mixing process has many industrial applications. When two fluids are the same kind but, each having different temperatures or other characteristics, the mix is calculated using the "lever" rule.

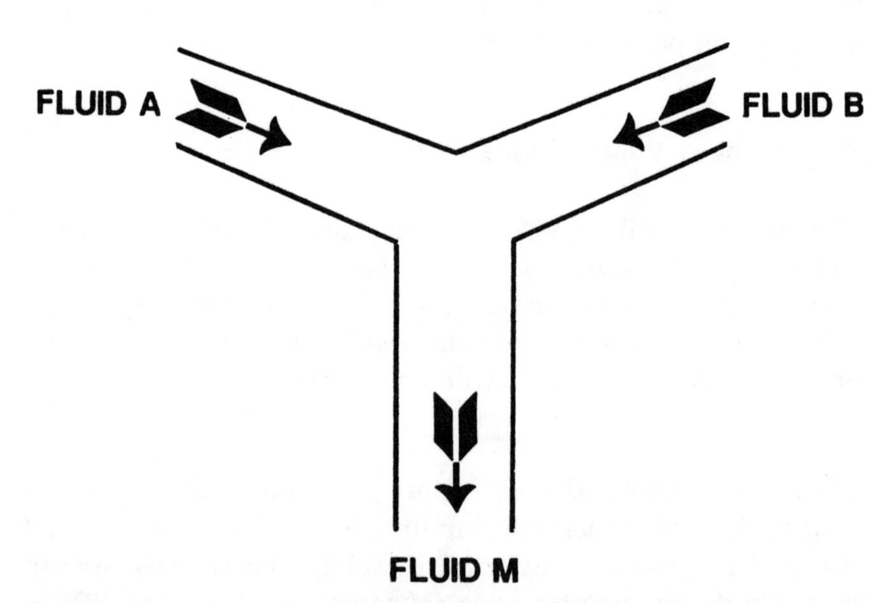

Figure 8-1. Continuous Mixing Process.

$m_A/m_B = (t - t_B) / (t_A - t)$

where:

m_A and m_B = Mass of fluids A and B, respectively
t_A = Temperature of fluid A
t_B = Temperature of fluid B
t = Temperature of mixture (fluid M)

An example of such mixing occurs in the mixing chamber of an air-handling unit, where the return air at 25 °C mixes with outdoor air at 5 °C as shown in Figure 8-2.

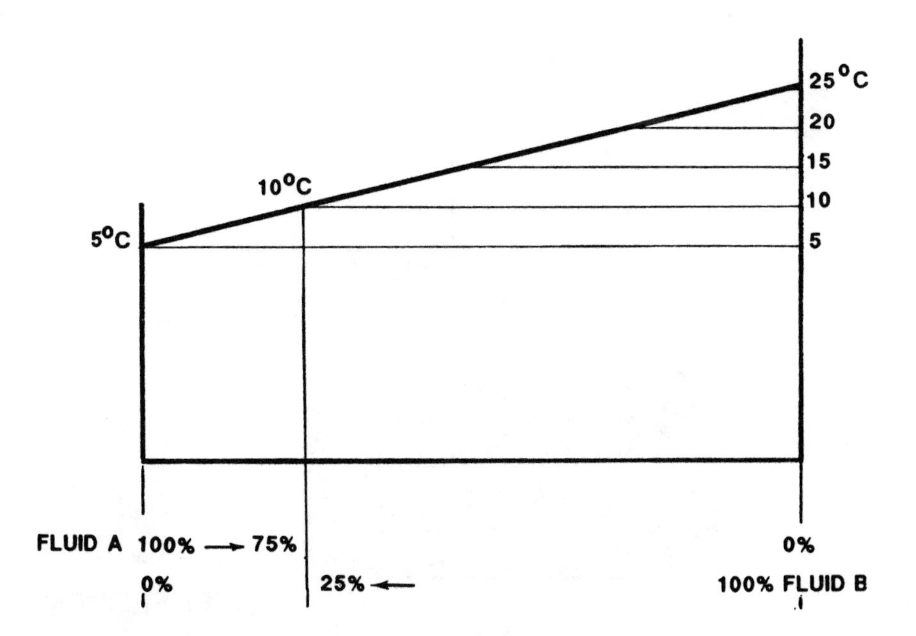

Figure 8-2. The Lever Rule of Mixing.

In this example, the question may be posed in one of two ways: What is the temperature of mixed air containing a certain percentage of outdoor air, or how much outdoor air is needed to maintain a certain mixed air temperature?

The most complex mixing process is time-dependent. Such a process goes through a transient state in which the concentration changes continuously until it reaches an asymptotic steady state level of mixing. This occurs when two fluids flow into a reservoir that is initially filled with one of the fluids and from which the mixture either flows, is exhausted or pumped out (refer to Figure 8-3).

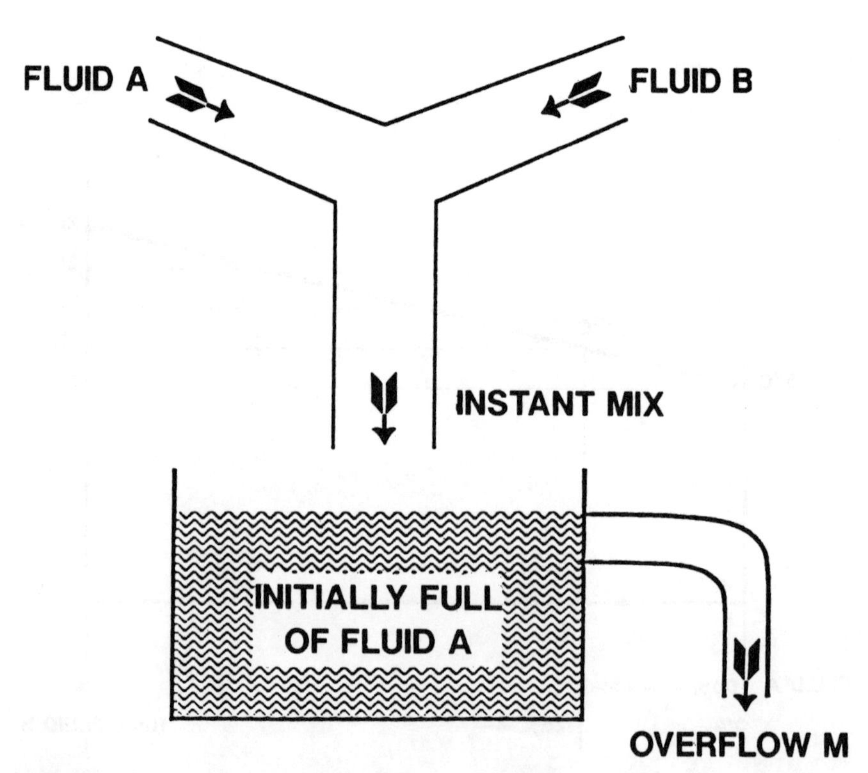

Figure 8-3. Mixing Process in a Reservoir.

This process is mathematically described by a differential equation. When solved, the equation describes the concentration of the mixed fluid in a set of solutions, the loci of which falls on an exponential curve which, after a substantial period of time, approaches the steady state concentration asymptote (refer to Figure 8-4).

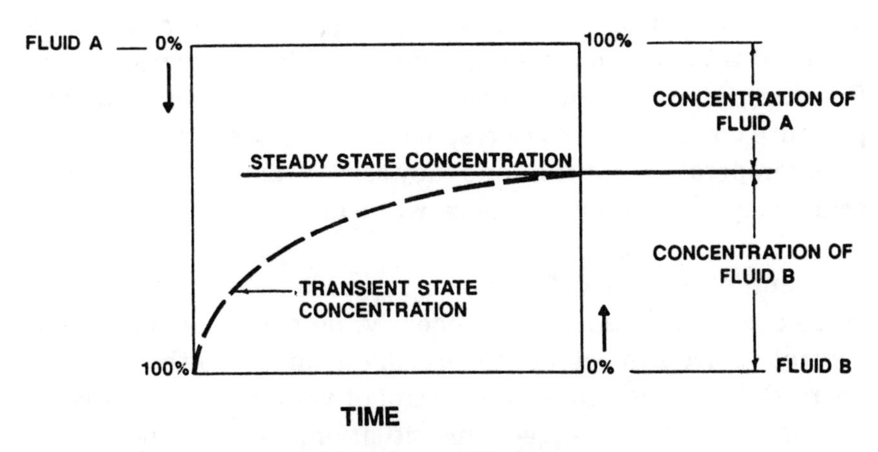

Figure 8-4. Time-Dependent Mixing.

The theory of mixing described above is based on an assumption that two fluids mix perfectly. Let us now apply this theory to the control of IAQ by general ventilation.

8.3 General Ventilation

In buildings, each indoor air contaminant is generated at its own rate while outdoor air is introduced at a specified rate. Only the most complex of mixing (time-dependent mixing) reflects the general ventilation process, however, with some special qualifications. While the dilution of indoor air contaminants is simultaneous, the concentration of each indoor air contaminant can be determined independently.

In office buildings (commercial and institutional), the general ventilation process is accomplished by air-handling systems. These systems can be classified into two categories depending on they achieve their effect whether modulating the outdoor airflow or providing a constant-volume of outdoor air. Systems in the first category are equipped with a free-cooling-cycle capable of providing cooling by mixing of return air with outdoor air when the outdoor temperature is low. There are several variations of air-handling systems that belong to this category. However, there are only a few that belong to the second category. The most commonly used system of this second category is the compartmental variable-volume system.

The primary function of air-handling systems is to deliver the required volume of air to the location at a desired temperature. For that purpose, air-handling systems are equipped with temperature control devices to respond to varying thermal loads in each temperature control zone they serve. Temperature control zones may contain one or more workstations.

The ventilation control function is superimposed on the temperature control function. Until recently, no real-time ventilation control device has been commercially available to the building operators and, therefore, the control of ventilation is relatively less understood. To correct this situation, first it is necessary to improve our understanding of the ventilation function not only among the building operators but also among the designers, contractors, and the equipment suppliers. For that purpose, it is necessary to examine the ventilation process in a single workstation using the concept of critical workstation. A critical workstation in a specified ventilation zone is the one with the most demanding ventilation requirements. Therefore, satisfying the ventilation requirements in a critical workstation implies that the ventilation requirements in all other workstations in the same ventilation zone will be also met or exceeded.

A ventilation zone usually consists of several workstations with varying ventilation parameters. It is, therefore, necessary to establish a method of correlating the ventilation performance in all workstations in a ventilation zone. This method involves a correlation of several mixing processes taking place simultaneously. One mixing process takes place at the mixing chamber, another in each workstation, and yet another in the return airstream. The correlation between ventilation in a critical workstation and the entire ventilation zone is mathematically expressed by the following equation in ASHRAE Standard 62-1989.

$$V_{ot}/\sum V_{st} = [\sum V_{on}/\sum V_{st}] \;/\; [1 + \sum V_{on}/\sum V_{st} - V_{oc}/V_{sc}]$$

The physical meaning of this equation is as follows: The critical workstation is supplied with air which is a mixture of the return and outdoor air (refer to Figure 8-5). This mixture is capable of diluting the indoor air contaminants generated in the critical workstation. This dilution capability of the mixture is also equal to the dilution capability of outdoor air supplied at a specified rate. This mixture contains a certain percentage of outdoor air.

The same mixture is also distributed to the noncritical workstations. Since these workstations are supplied with air mixture based on the need of the critical workstation, they receive more outdoor air than required for dilution. Consequently, return air from the noncritical workstations retains some unused dilution capability.

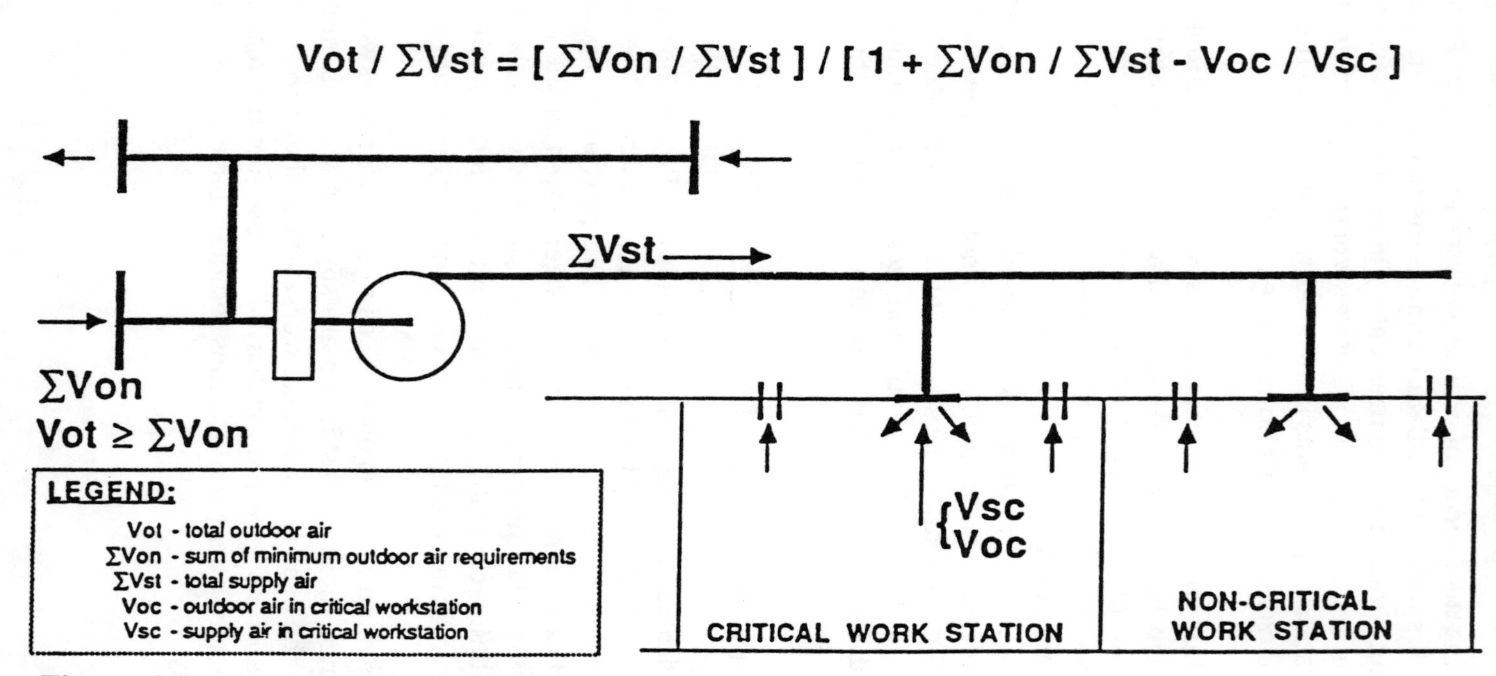

Figure 8-5. Multi-Workstation Ventilation Process.

Return air from all workstations (critical and noncritical) is mixed when returning to the air-handling unit. The unused dilution capability in this air is augmented by sufficient outdoor air to provide the dilution capability necessary to control IAQ in the critical workstation.

The relationship between ventilation of a single workstation and that of the entire ventilation zone is one of the fundamental pieces of information necessary for the control of ventilation rates. The other is the established rate of CO_2 generation by the occupants. With this information available, the carbon dioxide method of general ventilation control becomes practical. The method is aimed at controlling air quality in offices, commercial, and institutional buildings.

8.4 The Carbon Dioxide Method of General Ventilation Control

An average person generates CO_2 at 0.0050 liter per second (L/s) in a sedentary activity such as those in offices, 0.0036 L/s at rest or sleeping, and 0.0072 L/s in physical activities typical of industrial environments. The rate of CO_2 generation varies from one person to another, also in one person from time to time depending on the changes in his metabolic rate. Although this correlation may seem to be very complex to represent the ventilation needs of all people at all times, it in fact offers an excellent response to the ventilation needs. With the CO_2 rate per person known and the outdoor air per person specified, the concentration of CO_2 can be determined as a function of time and be used as a calculated setpoint in a typical ventilation control loop. In a ventilation control loop, an actual or a measured concentration is compared with that predicted. When these two concentrations agree, then the outdoor air damper is set correctly and needs no adjustments. When there is no agreement between the two concentrations, an appropriate adjustment of the damper must be made. This represents the concept of IAQ control based on controlling ventilation rates using CO_2 concentration as a surrogate.

8.4.1 Ventilation Variables

The algorithm used in determining the control setpoint is demonstrated in the following example in an office building ventilating zone. The following calculations are performed by the computer program CO2.BAS. The program is in Basic and uses imperial units only. Following are the ventilation variables used:

Occupied area	100,000 sq ft
Ceiling height	11 ft
Number of people present	500
Critical workstation size	120 sq ft
Air supply to critical workstation	90 cfm
Outdoor air per person	20 cfm
Total air supply to ventilation zone	120,000 cfm
Free-cooling cycle	yes
Return air temperature	75 °F
Supply air temperature	55 °F
Outdoor air temperature	90 °F
Outdoor CO_2 concentration	340 ppm
Initial CO_2 concentration	350 ppm

It should be noted that if the ceiling space is used as a return air plenum, its volume must form a part of the mixing reservoir. Therefore, 11 ft rather than the usual ceiling height of 9 ft or 9.5 ft was used. The calculated CO_2 concentrations are graphically presented in Figure 8-6.

The CO2.BAS is also capable of analyzing the effect of changes in each of the variables. This is accomplished by selecting the analyzed variable from a menu, assigning the minimum and maximum limits of the investigative range, and the steps at which the calculations are to be performed within the specified range. In this example, the concentration of CO_2 in steady state, air changes per hour, and the average outdoor air supply per person were calculated for each step between the minimum and maximum limits. The last calculated output provides a measure of overventilation in the ventilation zone. Table 8-1 shows the results of an analysis in which the effect of outdoor air per person is examined.

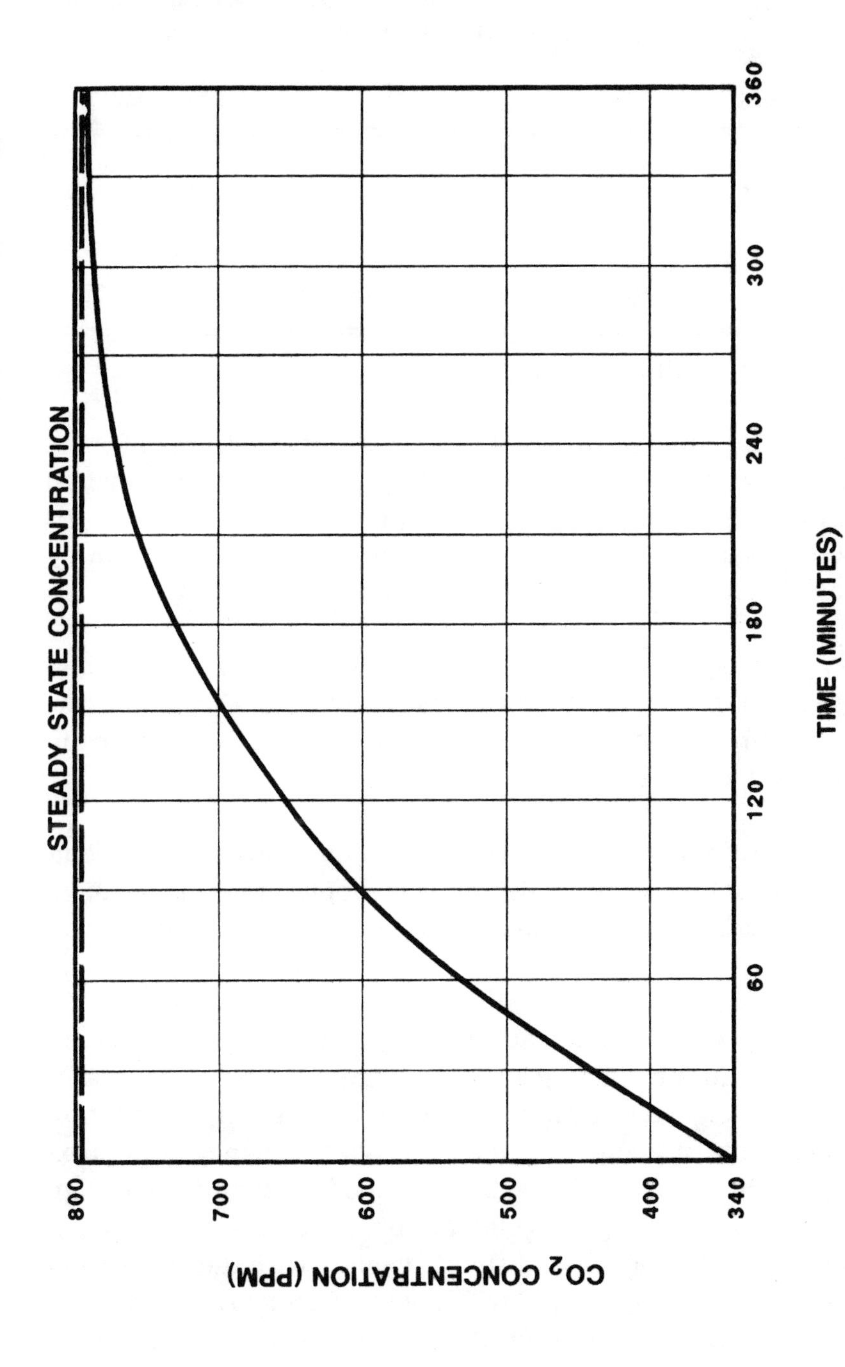

Figure 8-6. Loci of Calculated Control Setpoints.

Table 8-1. Impact of Outdoor Air Supply on Ventilation.

Specified Outdoor Air Supply (cfm/person)[A]	Steady State CO_2 Concentration (ppm)	Air Changes per Hour (hr^{-1})	Average Outdoor Air Supply (cfm/person)
5	2385	0.14	5
10	1326	0.29	11
15	973	0.46	17
20	796	0.63	23

A: Delivered to critical workstation.

The output shown in Table 8-1 indicates that the degree of over-ventilation at a ventilation rate of 20 cfm per person is 23/20 and that the concentration of 1000 ppm specified in the ventilation standards can be achieved in the critical workstation by a ventilation rate of 15 cfm per person.

The most frequently asked questions with respect to this method are related to the effects of variable-air-volume (VAV) and variable occupancy. Answers to these questions are evident from Tables 8-2 and 8-3.

The variation in CO_2 concentration in steady state can be compensated for by measuring the total air supply and using the measured signal to make adjustments to the calculated control setpoint.

The steady state concentration varies slightly with changes in occupancy. In the building examined, the make-up air for washroom exhaust provides enough outdoor air to satisfy ventilation requirements of about 60% of the design occupancy of approximately 500 people. In Table 8-3, it is evident that if the design occupancy was changed to 300, or 60% of the actual design occupancy, the method would always comply with the ventilation requirements. This strategy will have a built-in error of providing an average of 24 cfm of outdoor air, rather than 23 cfm.

Table 8-2. The Effect of Variable-Air-Volume on Ventilation.

Total Air Supply (cfm)	Steady State CO_2 Concentration (ppm)	Air Changes per Hour (hr^{-1})	Average Outdoor Air Supply (cfm/person)
50,000	858	0.56	20
60,000	840	0.58	21
70,000	828	0.59	22
80,000	818	0.60	22
90,000	811	0.61	22
100,000	805	0.62	23
110,000	800	0.63	23
120,000	796	0.63	23
130,000	793	0.64	23
140,000	790	0.64	24

Table 8-3. The Effect of Variable Occupancy on Ventilation.

People Present	Steady State CO_2 Concentration (ppm)	Air Changes per Hour (hr^{-1})	Average Outdoor Air Supply (cfm)
100	761	0.14	25
200	770	0.27	25
300	778	0.40	24
400	787	0.52	24
500	796	0.63	23
600	805	0.75	23

Another variable to observe is the outdoor air temperature. Figure 8-7 shows the air changes per hour (ach) as a function of outdoor air temperature. Notice that the ventilation rate increases from 0.5 ach in the mechanical cooling mode to 4.0 ach when the system is switched to the free-cooling cycle at the outdoor air temperature of 55 °F.

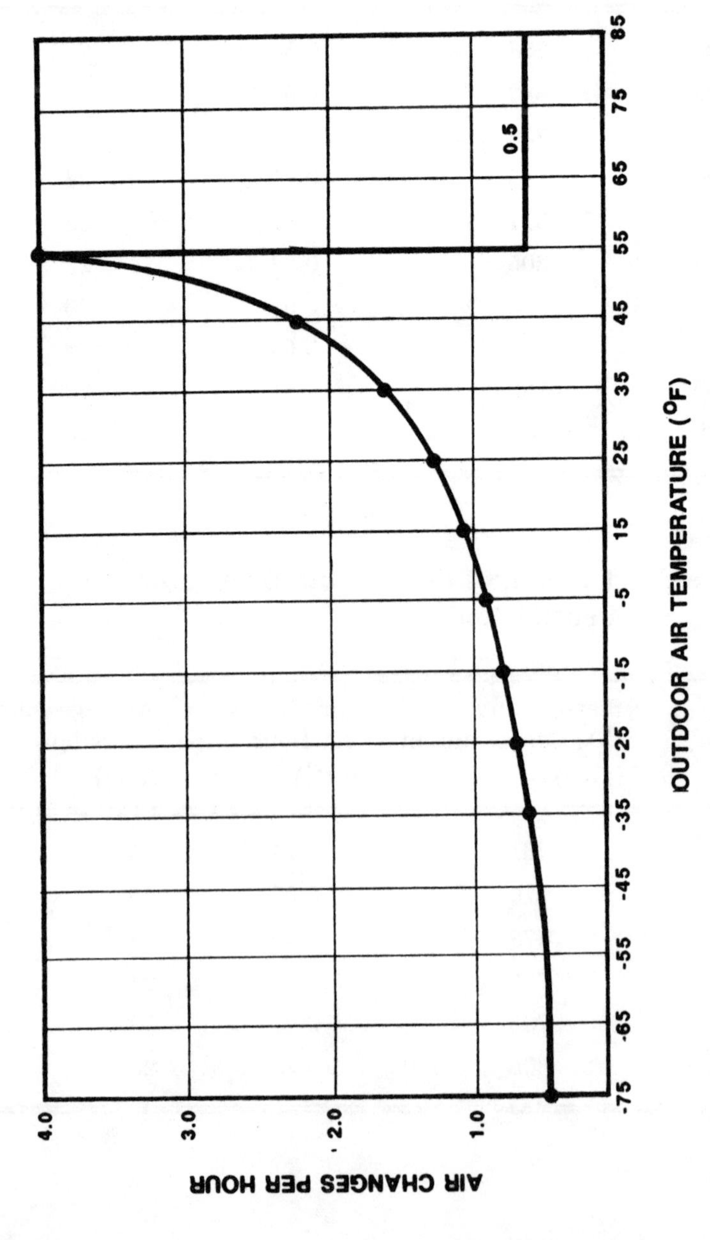

Figure 8-7. Ventilation Rates in a Free-Cooling Cycle.

The outdoor concentration of CO_2 also affects the calculated setpoint, but is easy to compensate for. The concentration in a control setpoint is directly proportional to the change in outdoor CO_2 concentration. For example, when the outdoor concentration increases by 50 ppm, the concentration in every calculated control setpoint also increases by 50 ppm. Measurement of outdoor CO_2 concentration is recommended as an input into the control algorithm.

Changes in metabolism are sensed by this method as changes to occupancy. The method recognizes how much CO_2 is generated, not how many people it is generated by. This is actually an advantage; people who work harder generate more CO_2 and, therefore obtain more outdoor air.

8.4.2 Ventilation Effectiveness

The above discussion did not involve the dilution process within a workstation. However, the effectiveness of the ventilation depends very much on local mixing of indoor air contaminants. Normally, the mixture should be obtained by supply air entering the workstation at such velocity and angle that sufficient mixing will take place at the reference point of the workstation which should be the location of the occupant. This, however, is not easily accomplished in all workstations at all times, especially if the air is supplied by a VAV system. In any event, a proportion of the supply air bypasses the reference point in the workstation. This results in a decreased ventilation efficiency, which effectively is the same as a reduced ventilation rate introduced into the workstation with a perfect ventilation efficiency.

Ventilation efficiency and its effects on the ventilation rates require further study. A practical solution which can produce higher ventilation efficiency involves the use of a local air mixing device such as a fan. The fan can be either standing on the desk, or incorporated into the furniture, desk lamp, or perhaps even be in the personal computer.

8.4.3 Control Strategy

The number of people present is the most important variable of the ventilation control algorithm and is hardest to determine. Since it is not practical to measure this variable automatically, it is necessary to use a strategy that employs another indicator as a surrogate measure of occupancy. The built-in error of such strategy should be as small as possible.

The following is a control strategy that complies with these conditions. It is based on observations of the change in the CO_2 concentration. When the change reaches a certain level, the control process is initiated. This overcomes the guesswork relating ventilation control to occupancy, especially in buildings with flexible hours of occupancy. The sequence of operation is as follows:

(a) When the washroom exhaust (and/or other) fan is turned on, the supply fan is also turned on. The return air damper remains shut and the outdoor air damper is set to its minimum to allow for make-up of the exhausted air. At that time, the calculations of the control setpoints for design occupancy and the measurement of CO_2 concentration in return and outdoor air are initiated.

(b) Calculate the degree of change in CO_2 concentration measured in the return airstream.

(c) Compare the changes measured in concentration with changes in predicted concentration at point $t_1 = t/4$ (refer to Figure 8-8). This point is arbitrarily selected. The selection is based on an in-depth analysis of variable occupancy levels. The operator should be able to make changes to the point in time with which the comparison is made, within a range of t/2 to t/8.

(d) When the change measured is larger than the calculated change, activate the control loop that controls the return air and outdoor air dampers. Follow the rate of change (the

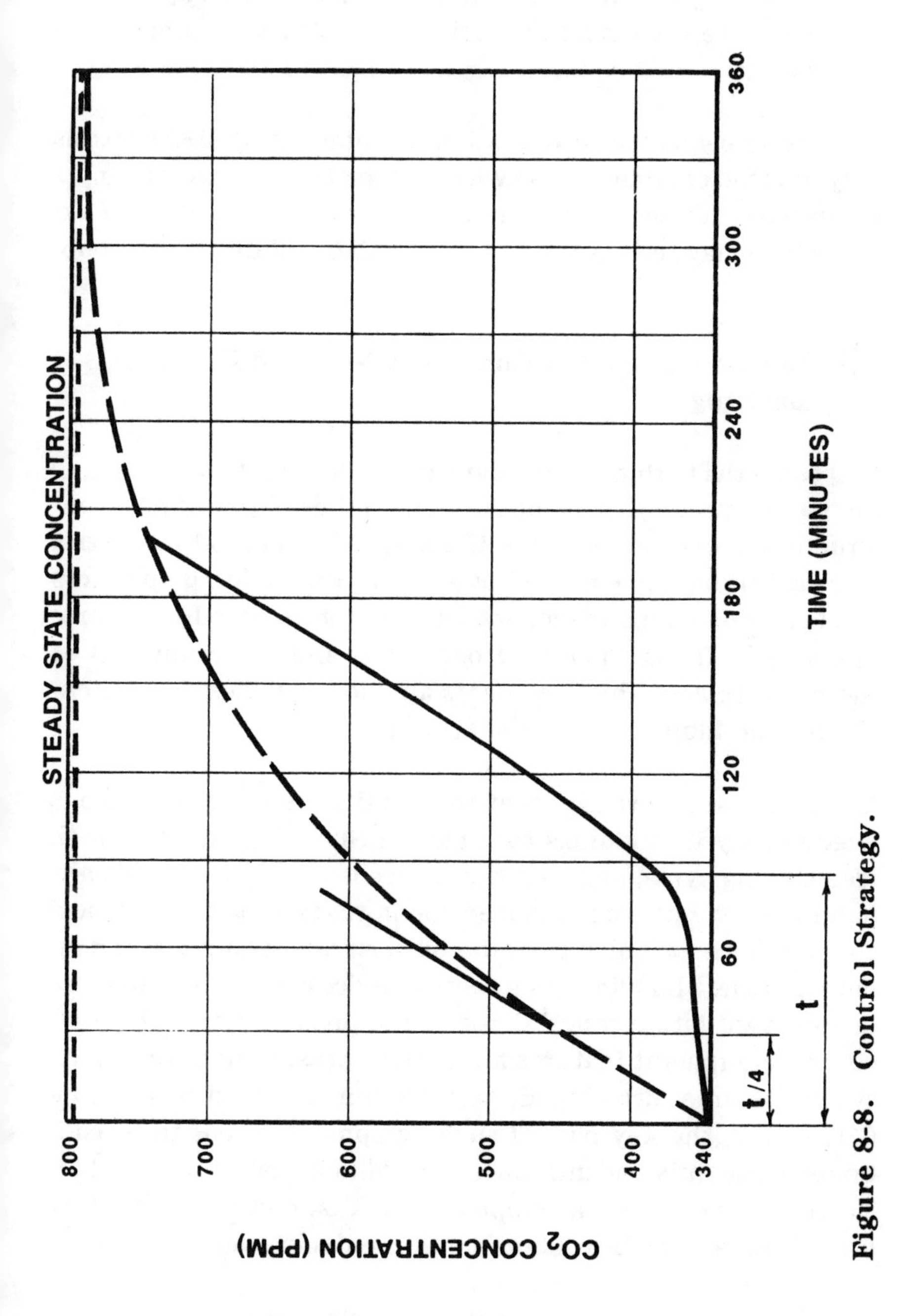

Figure 8-8. Control Strategy.

angle of concentration change) until the straight line, representing this rate of change, intersects the exponential curve representing the loci of the calculated control setpoints.

The above sequence of operation is suitable for DDC controls only; it is too complex to use with pneumatic controls. Attempts to simplify this sequence may result in the introduction of errors which may seriously affect the reliability of the control loop.

8.5 Indoor Air Quality Control in New and Existing Buildings.

IAQ concerns both the new and existing buildings. The existing number of buildings in comparison with the volume of new construction is enormous. Since the suggested method to control IAQ cannot be implemented over night, the building operators must formulate an implementation plan for its introduction providing both for the needs of occupants and the economics of the marketplace. This represents an enormous business potential for the industry in the next decade.

It is important to realize that the ventilation control device is needed only in buildings that have air-handling systems with free-cooling cycle, and that the device is needed when the air-handling systems are not in free-cooling cycle but in mechanical cooling mode at which time the outdoor supply air is at a minimum. In these buildings, one control device can serve up to three or four ventilation zones. Some of the presently available CO_2 control equipment features multipoint measurement through a system of tubes drawing air samples from selected points. The CO_2 sensor, the key part of such equipment, needs to be calibrated frequently and introduces an additional labor cost in O&M. However, the latest developments in CO_2 control technology provide for automatic calibration, thus reducing the cost of O&M.

In buildings with compartmental variable-volume systems, the supply of outdoor air is constant and limited. To provide for possible maximum satisfaction with the limited constant outdoor air supply, it is important to ensure that the system has the capability to deliver the outdoor air where the occupants are. In such cases the ventilation control device may also be employed, but in a different way. The devices that control the supply of outdoor air to on-floor fan rooms are modulated to keep the CO_2 concentration uniform throughout the entire building.

Some of the buildings with older compartmental systems were designed at a time when the required ventilation rates were considerably lower than the ones listed in the current ventilation standards. These buildings may not have the capability to comply with the current ventilation rates on a real-time basis. In these buildings, compliance must be on the basis of air changes per day (refer to Figure 8-9).

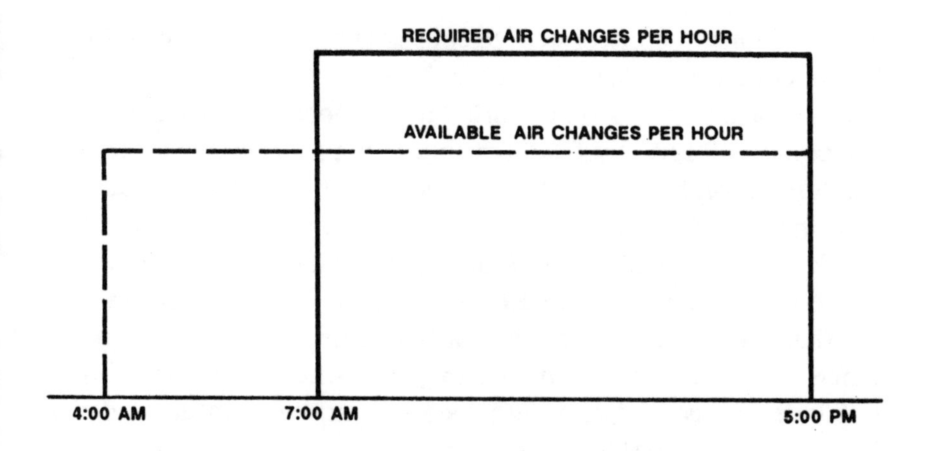

Figure 8-9. Alternative Ventilation Schedule.

These ventilation systems need to be started early in the day to produce the required number of air changes per day at the end of the occupancy period. Additional research is needed to prove that this operation strategy offers the same, if not better, performance of the general ventilation systems.

8.5.1 Control of Out-Gassing Products

The ventilation standards are intended to control all indoor air pollutants generated in mature offices, institutional, and commercial buildings. However, when the building is newly furnished, the out-gassing process may be stronger than expected. Where a predominant out-gassing indoor air contaminant can be identified, the rate of generation can be measured by the same mixing process used to determine the air changes per hour necessary to keep that air contaminant under control. It is a good practice to start ventilation systems in time to complete one air change before the start of occupancy on days following a regular workday, and two air changes on days following weekends or holidays. More accurately, a set control strategy must be based on known generation rates of the contaminants and on known decay of their concentrations in time.

8.6 Commissioning and Verification

The ventilation control device senses the concentration of CO_2 in the common return airstream. The correlation with the CO_2 concentration in a critical workstation needs to be verified during commissioning by field testing. It is proposed that the CO_2 concentration be recorded in all workstations identified as critical and adjustments made to the main algorithm to the computer program CO2.BAS if any substandard performance is detected. If there is a need to use more refined procedures, an artificial source of CO_2 such as a fire extinguisher can be used. Since CO_2 is not toxic, no extra precaution for the testing is needed. The commissioning process is explored in great detail in Chapter 14, Part 3.

8.7 Further Research

IAQ control requires further research in several areas. One of them is the research related to the refinement of the general ventilation control method. Such work will include improved reliability, improved understanding of ventilation performance and its effect on ventilation rates, improved procedures, new instrumentation to control specific air contaminants by the general ventilation process, etc.

When the ventilation control device is installed in enough buildings, these can serve as laboratories for research based on statistically treated opinion surveys. With a known degree of compliance with the ventilation standards, serious work can begin on correlation of ventilation rates with the degree of acceptance. This work will hopefully produce a new IAQ standard that would be expressed in terms more suitable for the day-to-day operation of buildings.

9. TECHNIQUES FOR MODELING VENTILATION EFFICIENCY AND AIR DISTRIBUTION IN OCCUPIED SPACES

Ren Anderson
Technology Leader, Solar Energy Research Institute
Golden, Colorado

9.1 Introduction

Ventilation systems have traditionally been designed to provide odor control and thermal comfort assuming that the air in a building is perfectly mixed. Increased awareness of the potential health risks associated with indoor air pollutants has stimulated interest in improving our understanding of the factors which may reduce the ability of a ventilation system to control these pollutants. The quality of indoor air is determined by the ability of the ventilation system to directly supply ventilation air to a building and remove pollutants before they mix with the room air. These performance characteristics depend in general on the building geometry, pollutant source characteristics, ventilation, thermal stratification, duct location, and the type of diffuser used. If the air in a room is well mixed, then the pollutant concentration can be determined based on the room ventilation rate, pollutant source-strength, and the concentration in the supply air. If the air in a room is not well mixed, then knowledge of local variations in performance is also required.

Because it is impractical or impossible to remove all pollutant sources, the building ventilation system must be designed to provide an adequate balance between the ventilation rate and pollutant source at all occupied building locations over a wide range of operating conditions. It must be designed to account for the worst case rather than average conditions because the sensitivity of concentration to flow nonuniformities can produce localized areas with unacceptable high concentrations even if an acceptable average concentration for a building can be achieved at a given ventilation rate.

It is important to recognize that inadequate ventilation is only one of many potential contributors to indoor air quality (IAQ) problems. Figure 9-1 summarizes some of the major factors that have been identified from previous studies. These results are obtained from a study conducted by the National Institute of Occupational Safety and Health (NIOSH) in 446 buildings. Although ventilation is only one of several factors that must be considered, it is a relatively simple problem to solve in most cases. Therefore, it is essential that the conditions which contribute to ventilation be identified and resolved so that poor ventilation can be positively ruled out as a cause of poor IAQ in a building. Inadequate ventilation can result from poor supply of outside air to a building zone by the ventilation system and nonuniform distribution of ventilation air after it is delivered to a building zone. In either case, providing an adequate supply of ventilation air at the outside air intake does not automatically ensure that an adequate amount of air is supplied to building occupants.

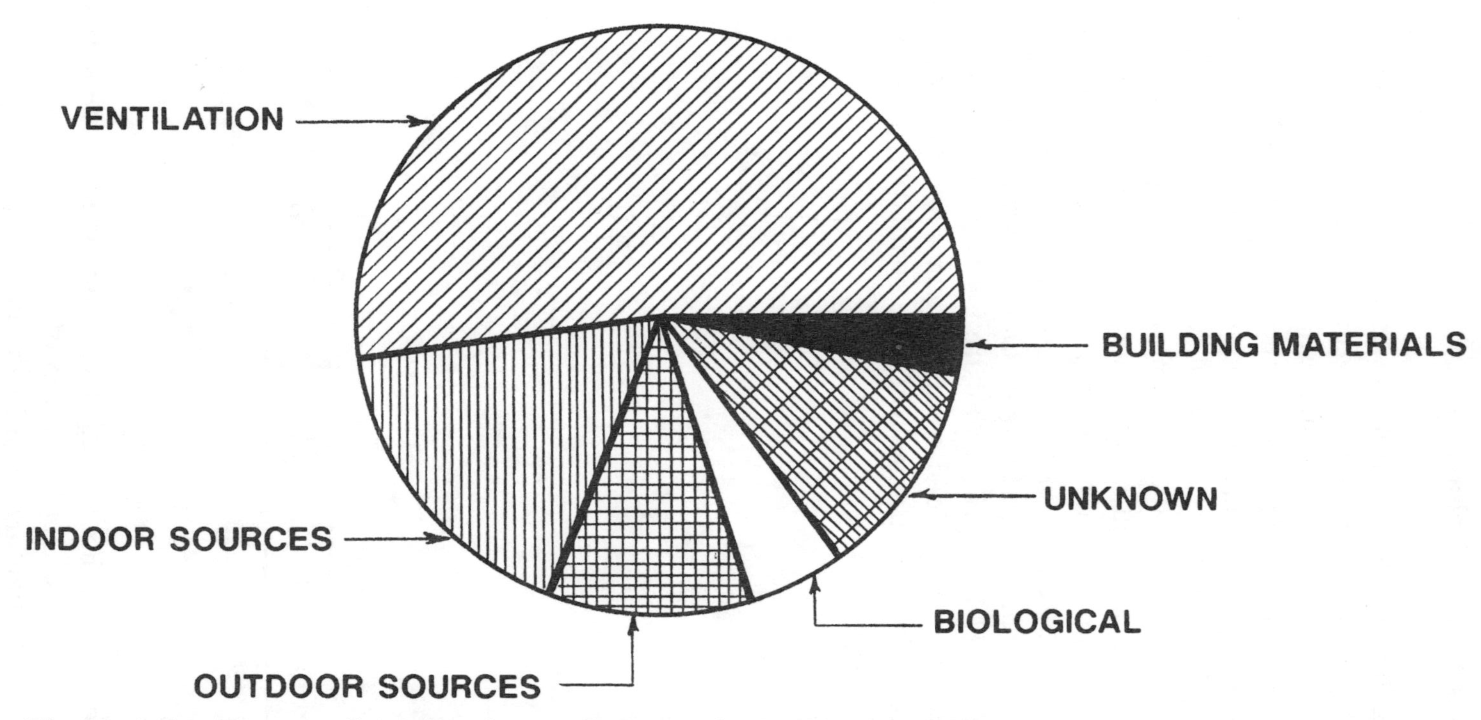

Figure 9-1. Factors Contributing to Indoor Air Quality Problems.

9.2 Diagnostic Techniques

Ventilation performance problems cannot be resolved unless the manufactures, engineers, building owners, and the code-making organizations have a clear understanding of the causes of poor performance and use reliable methods to diagnose performance problems. These organizations and their individual functions are identified in Figure 9-2. Several different ventilation efficiency measures have been proposed to measure ventilation system performance (1, 2). Point measurement techniques (3-6) shown in Figure 9-3 are based on the analysis of concentration time series and can be used to determine the characteristic concentration scales or time scales for ventilation flows. The control volume techniques (7-11) shown in Figure 9-4 are based on the mass balances on well-defined control volumes and can be used to provide direct measurements of room ventilation rates and pollutant removal rates.

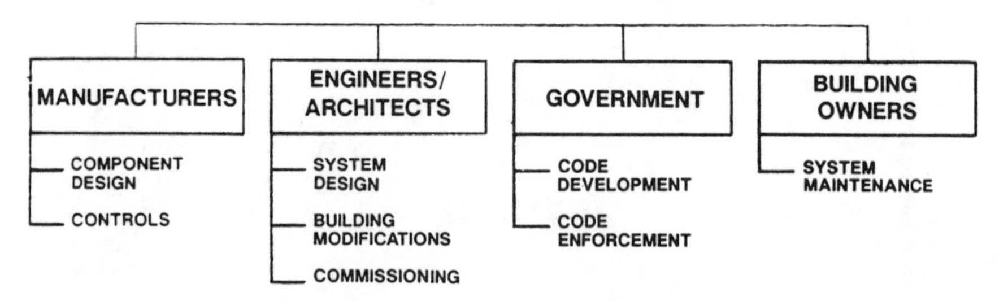

Figure 9-2. Major Organizations Involved in Identification and Solution of Ventilation Performance Problems.

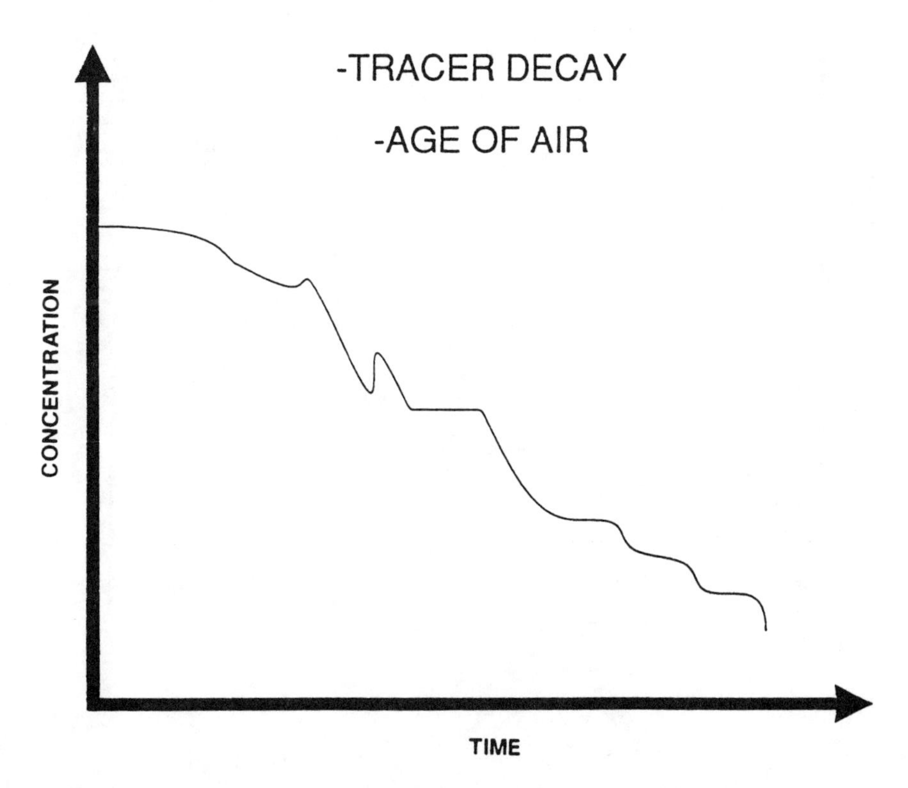

Figure 9-3. Diagnostic Techniques, Point Measurements.

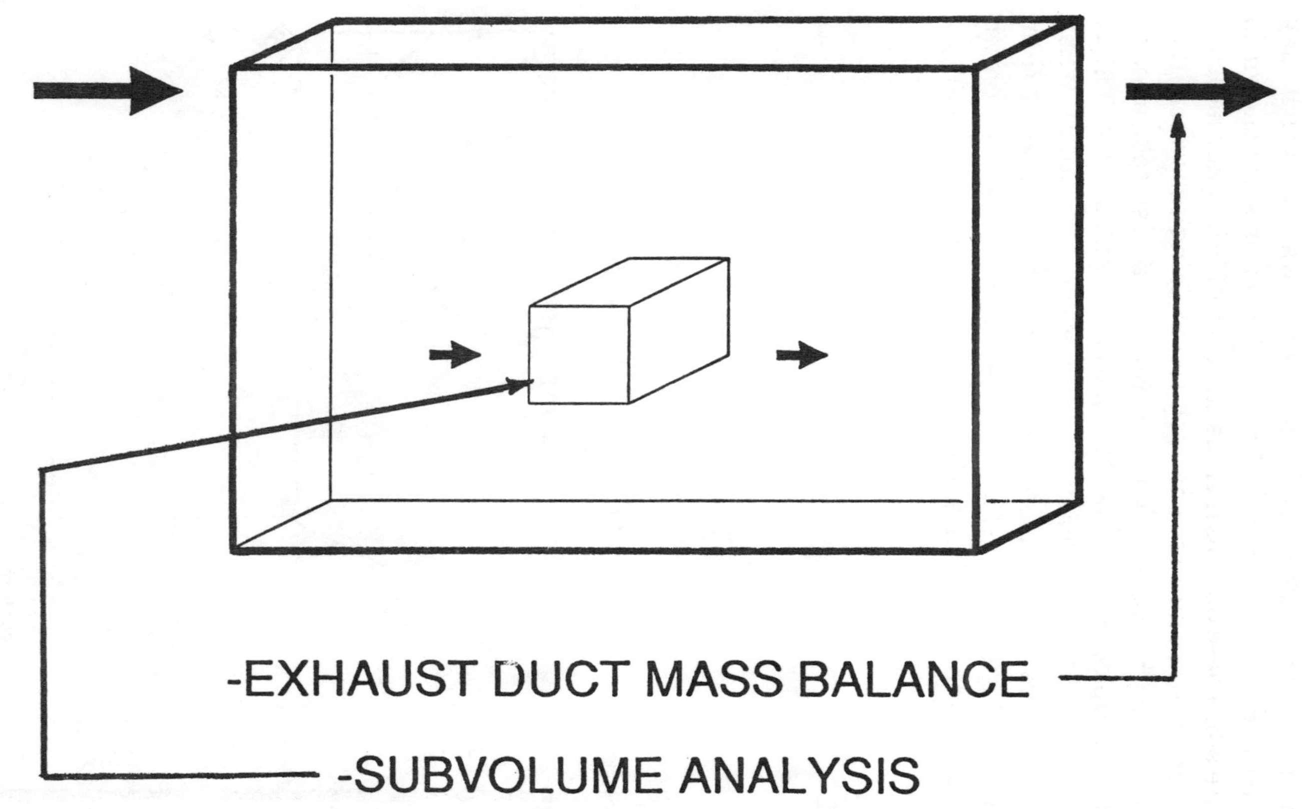

Figure 9-4. Diagnostic Techniques, Control Volume Measurements.

One of the main difficulties associated with the use of ventilation efficiency concepts is the number of ways in which they can be implemented. Some of the major options are shown in Figure 9-5. These options include the type of efficiency measure (air supply or pollutant removal), the building subsystem that the efficiency measure is applied to (distribution system, a room, or an occupied zone), and the measurement approach (test chamber, field measurement, or numerical calculation). When one considers the number of approaches that can result from combining Figures 9-3, 9-4 and 9-5, the difficulty associated with interpreting the results of different measurements becomes quite apparent.

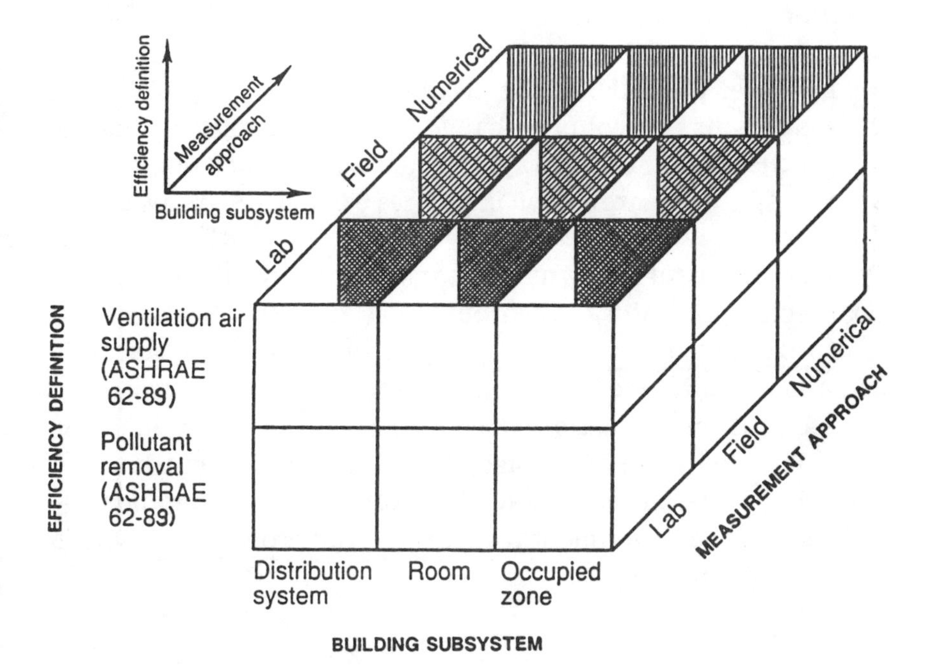

Figure 9-5. Implementation Options.

Test chambers provide a useful method of evaluating the effect of system design on ventilation performance because of the ability to vary test conditions and quickly determine the cause of poor performance. The level of complexity that can be studied in test chamber studies varies from simple small scale tests to room size tests using full scale heating, ventilating and air-conditioning (HVAC) system components. Full scale tests provide detailed information about the performance of specific components. Small scale tests provide benchmark data about simple test cases that can be used to compare different measurement techniques as well as to determine the parameters having the greatest effect on system performance. Small scale test chambers are designed on the same principles used in wind tunnel testing of buildings, aircraft, and automobile models.

Examples of ventilation air delivery rates measured at Solar Energy Research Institute (SERI) of Golden, Colorado using small scale testing techniques are shown in Figures 9-6 through 9-8. The supply duct and return duct consist of single slots which extend across the entire width of the test chamber. Flow recirculation causes air delivery to vary between 40% and 70% in the lower region of the test chamber. A value of 0.5 in Figure 9-6 indicates that 50% of the air at that location is replaced by ventilation air per volume change of supply air. A topological plot of the data and cutaway view showing details of the dead zone in Figure 9-6, are shown in Figures 9-7 and 9-8, respectively.

The measurements shown in Figures 9-6 through 9-8 use an optical technique (refer to Figure 9-9) based on the measurement of changes in light intensity due to absorption by pollutant tracers, and are analyzed using control volume methods. Digital image analysis techniques are used to speed the evaluation of the data. The digitizing board uses RGB input with eight bits per color and a spatial resolution of 756x486 pixels. A detailed description of the experimental apparatus and procedure has been previously reported (9).

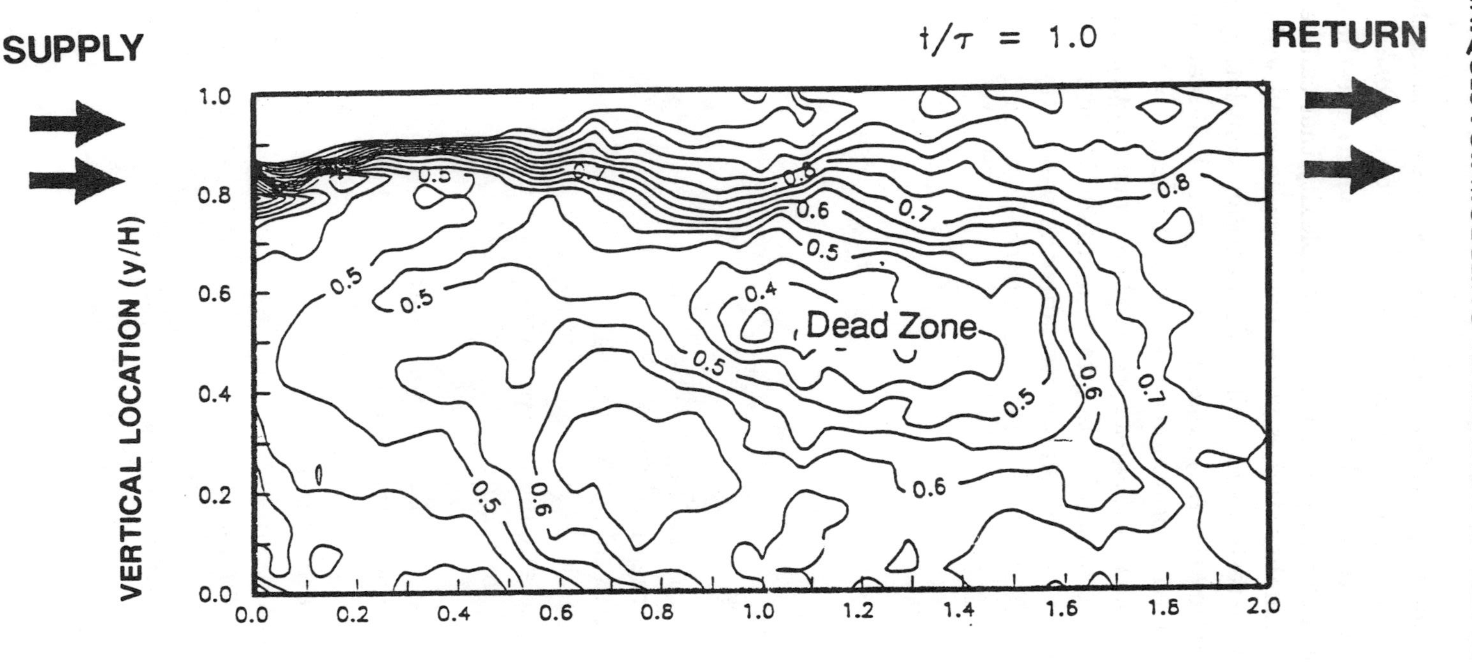

Figure 9-6. Small Scale Measurements of Ventilation Air Delivery.

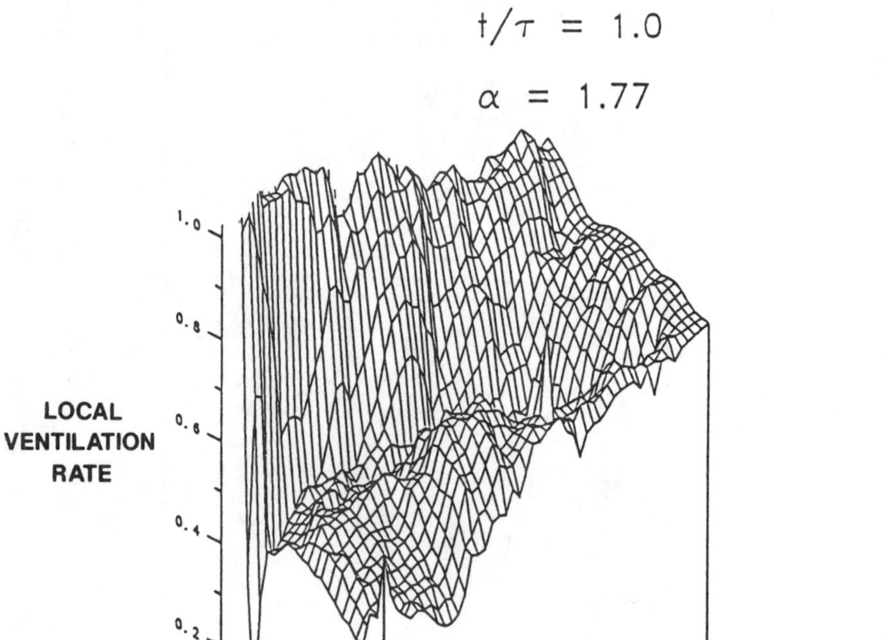

Figure 9-7. Topological Plot of Data in Figure 9-6.

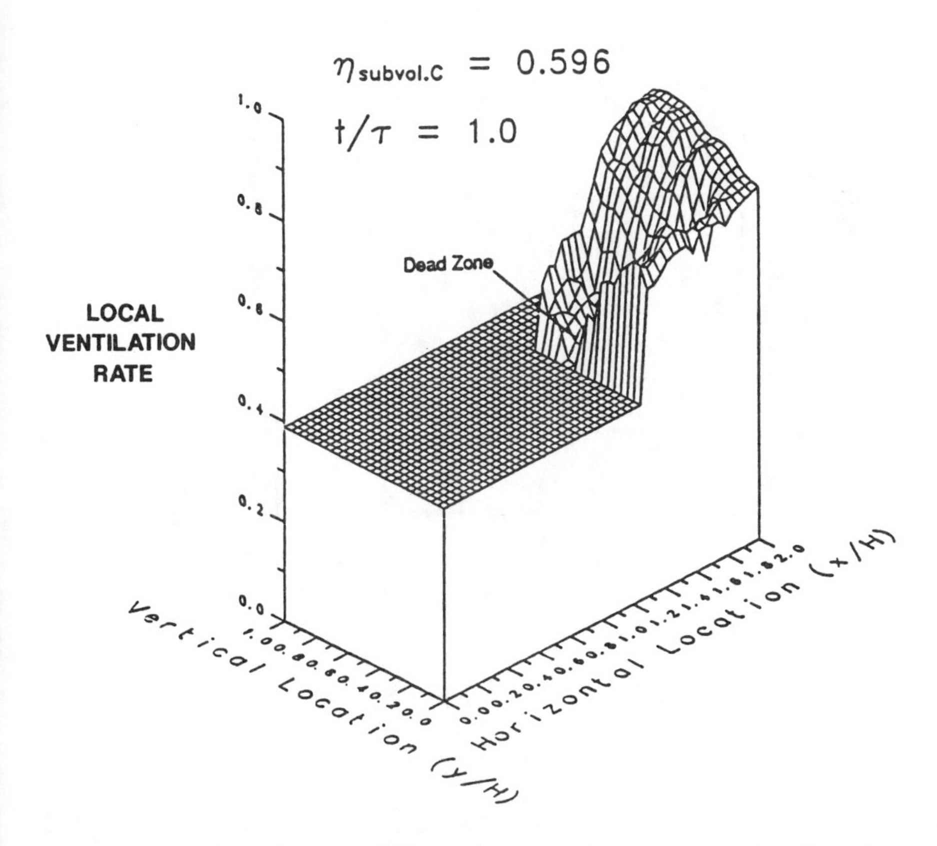

Figure 9-8. Cutaway View Showing Details of the Dead Zone in Figure 9-6.

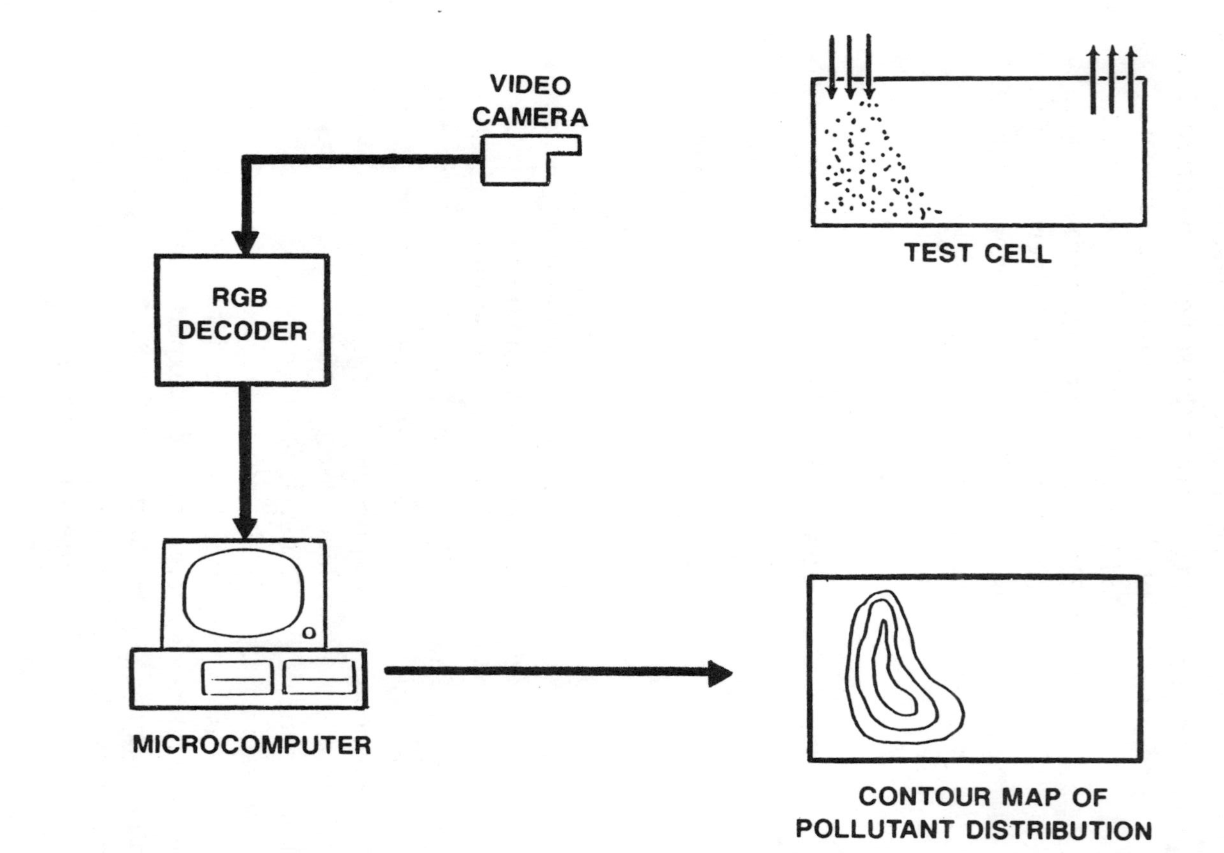

Figure 9-9. Ventilation Performance Measurement Using Digital Image Analysis Techniques.

Tests using the simple benchmark geometry shown in Figure 9-6 indicate that supply temperature has a large effect on ventilation performance. Representative results for heating (supply hotter than room), no load (supply at the same temperature as the room), and cooling (supply colder than room) are shown in Figure 9-10. Also included in Figure 9-10 are the field test results previously reported (7). Because a perfectly mixed system has an air delivery efficiency of 63% (10), the delivery performance in cooling applications is actually higher than that produced by perfectly mixed systems as shown in Figure 9-10. Conversely, the air delivery is reduced during heating conditions. Percent air delivery can be increased under all load conditions by increasing the recirculation rate as shown in Figure 9-11. However, this increase is achieved by reducing the rate at which outside air is supplied to the building.

9.3 Conclusions

Inadequate ventilation rate is one of the factors that can contribute to IAQ problems. Low ventilation rates can be a result of poor system design and operation, and poor room air distribution. The large number of techniques and implementation options for ventilation efficiency measurements make it difficult to compare measurements made by different research groups. Simple benchmark tests provide an important basis for such comparisons and also provide a testing ground for development and improvement of ventilation performance measurements. Benchmark tests with simple slot diffusers indicate that supply temperature is an important factor in determining ventilation delivery rates. Perimeter zones that experience large variations in heating and cooling loads as a function of time of day and season must be carefully designed to avoid correspondingly large variations in ventilation performance.

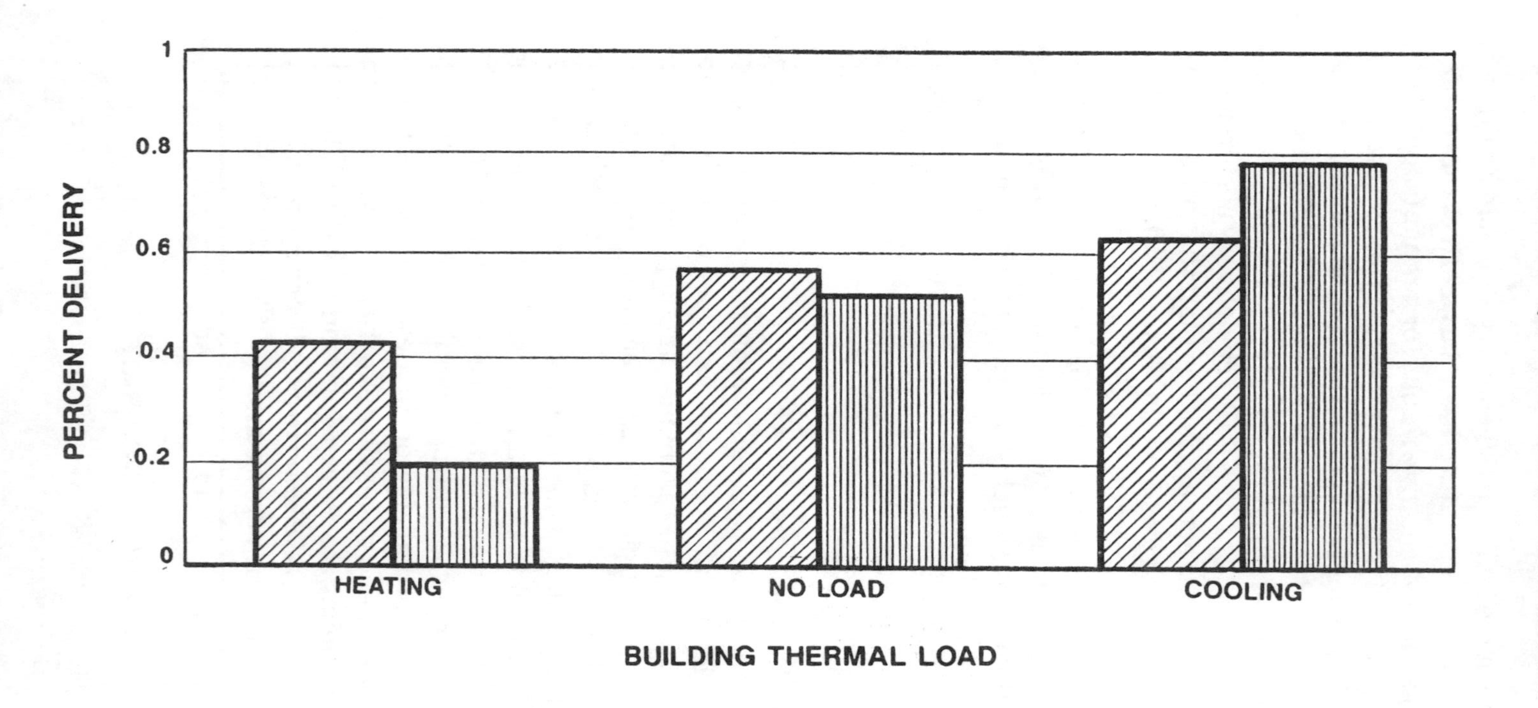

Figure 9-10. Impact of Supply Temperature on Air Delivery (100% Outside Air).

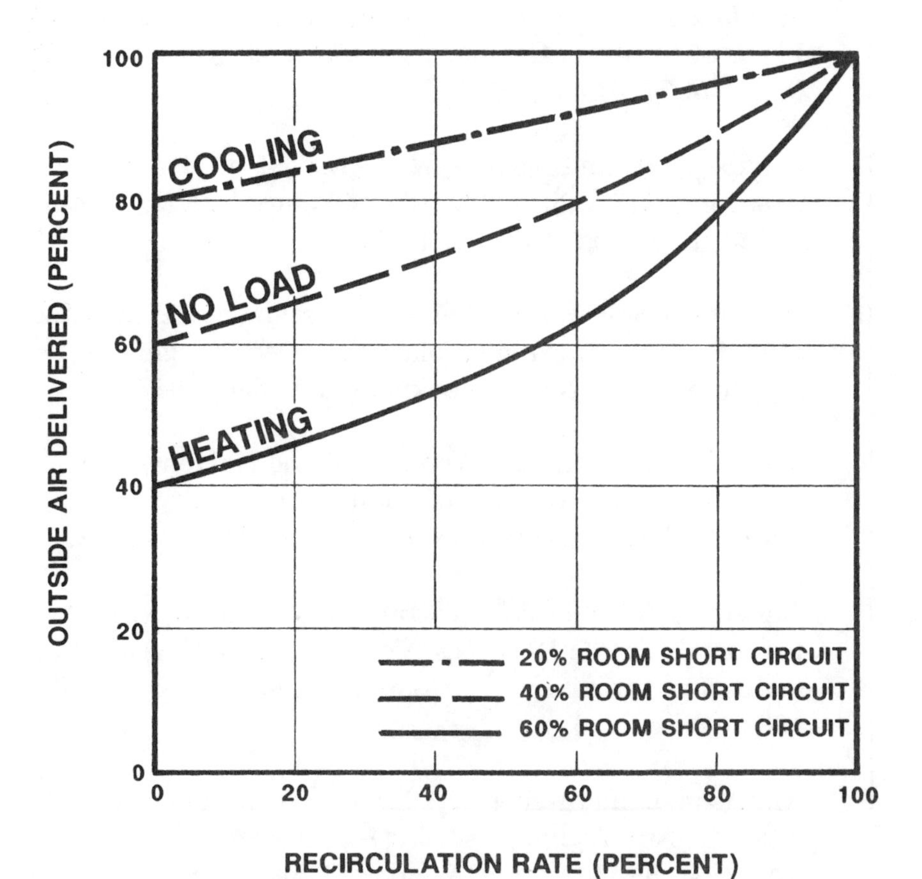

Figure 9-11. Impact of System Recirculation on Air Delivery.

References

1. Liddament, M.W. "A Review and Bibliography of Ventilation Effectiveness - Definitions, Measurements, Design and Calculation." Air Infiltration and Ventilation Centre, Technical Note 21, 1987.

2. Skaret, E. "A Survey of Concepts of Ventilation Effectiveness." SINTEF, Report STF15 A4057, 1984.

3. Danckwerts, P.V. "Local Residence Times in Continuous-Flow Systems." *Chemical Engineering Science*, 9, pp. 78-79, 1958.

4. Spalding, D.B. "A Note on Mean Residence Times in Steady Flows of Arbitrary Complexity." *Chem. Engineering Sci.*, 9, pp. 74-77, 1958.

5. Sandberg, M. and Sjoberg, M. "The Use of Moments for Assessing Air Quality in Ventilated Rooms." *Building and Environ.*, 18, pp. 181-197, 1983.

6. Rydberg, I. and Kulmar, E. "Ventilationens Effektivitet vid Olika Placeriingar av Inblasnings - Och Utsugningsoppningarna." VVS No. 3, Stockholm, Sweden, 1947.

7. Janssen, J.E.; Hill, T.J.; Wood, J.E.; and Maldonado, E.A. "Ventilation for Control of Indoor Air Quality: A Case Study." *Environ. International*, 8, pp. 489-496, 1982.

8. Janssen, J.E. "Ventilation Stratification and Air Mixing." Indoor Air Vol. 5: Buildings, Ventilation and Thermal Climate, Swedish Council for Building Research, pp. 43-48, 1984.

9. Anderson, R. and Mehos, M. "Evaluation of Indoor Air Pollutant Control Techniques Using Scale Experiments." ASHRAE IAQ '88, Atlanta, GA, pp. 193-208, 1988.

10. Anderson, R. "Determination of Ventilation Efficiency Based Upon Short Term Test." 9th AIVC Conference, Gent, Belgium, Sept., 1988.

11. Martin, D.; Anderson, R.; Farrington, R.; and Weaver, N. "Impact of Ventilation System Design on Exposure to Indoor Air Pollutants." EPA/AWMA Symposium on Measurement of Toxic and Related Air Pollutants, Rayleight, NC, May, 1989.

10. INDOOR AIR QUALITY SIMULATION BY USING COMPUTER MODELS

Howard Goodfellow
President, Goodfellow Consultants Inc.
Mississauga, Ontario, Canada

Page

10.1 Introduction

Indoor air quality (IAQ) may be defined as the nature of air that affects the health of individuals and their well-being. Air quality is also an indication of how well the indoor air satisfies the three requirements for occupancy: (a) thermal acceptability, (b) normal concentrations of oxygen and carbon dioxide, and (c) suppression of indoor air contaminants below their acceptable limits.

The development and application of validated mathematical and physical models provide the means to relate the design of a physical system to the expected exposure of occupants, related dose, and the resultant occupant response. Predicting the effectiveness of ventilation systems is difficult due to complex interaction of the system components within a ventilated space. Several numerical models developed to date have had varying degrees of sophistication, applicability, and accuracy. A brief review of the currently available numerical models is as follows:

Horstman (1) developed a numerical model that predicts the velocity distribution, airflow circulation patterns, and the airborne contamination distribution in an air exhaust passenger cabin. This model utilizes the finite difference approximations of the two-dimensional, time-varying Navier-Stokes equations. However, it is difficult to apply the model to other ventilated volumes, such as multi-cell volumes.

Meckler and Janssen (2) developed a model to calculate the amount of outdoor air required, concentrations of filtered contaminants, or the recirculation needed. This model is a single-cell, steady-state model which includes recirculation of the indoor air through filters.

Sparks et al. (3) developed a multi-compartment model based on a well-mixed mixing model. It estimates the effects of forced-ventilation, air-cleaning, room-to-room air movement, and natural ventilation on pollutant concentrations.

Crommelin and Buringh (4) developed a multiple-cell model to calculate the temperatures and concentrations of indoor air pollutants. They concluded that a mathematical model which calculates the airflows in an enclosure requires the numerical solution of the Navier-Stokes equations with a large number of grid points, which requires a large computational effort. They decided to simplify the model by dividing the enclosure into cells and estimating the airflow between cells from measurements taken in the enclosure.

Awbi (5) developed a numerical model (ROOMVENT) which solves two- and three-dimensional ventilation problems. The model solves the steady-state conservation equations of mass, momentum, and energy in finite difference form.

Siurna and Bragg (6) modeled room air diffusion based on a two-cell configuration which enables room air concentrations to be described stochastically. Through a stochastic analysis of a ventilation system, the model provides quantitative statistical conclusions concerning the ventilation performance.

Fletcher and Johnson (7) studied the buildup in concentration of a heavier than air gas emitted at various concentrations and positions within a room for different ventilation patterns.

Skäret (8) developed a ventilation model which is described in terms of the ventilation efficiency, and quantified by means of a two-zone (compartment) flow and difference model.

Dellagi et al. (9) developed a three-dimensional, time-varying ventilation model which calculates only the velocity profiles. It can also be adapted to calculate concentration profiles.

The model developed by Goodfellow Consultants Inc. (GCI) is a multi-cell, time-dependent model similar to the models of Crommelin and Buringh and Sparks et al. It has been expanded to include the multiple, time-varying emission rates of indoor air contaminants.

10.2 Theory of Modeling

10.2.1 A Fundamental Model

The single-compartment model as shown in Figure 10-1 includes various factors affecting indoor exposure: indoor and outdoor sources, removal, and dilution control. It identifies three most commonly used methods of control: (a) source control, (b) removal control, and (c) dilution control. Source control, N, may be represented by isolation, product substitution, or local exhaust. Removal control, E, may be represented by passive mechanisms such as settling or sorption, and active mechanisms such as fan filter modules, clean benches, or central forced-air systems with recirculated air. Dilution control, V_O, may be represented by infiltration, natural ventilation, or mechanical ventilation.

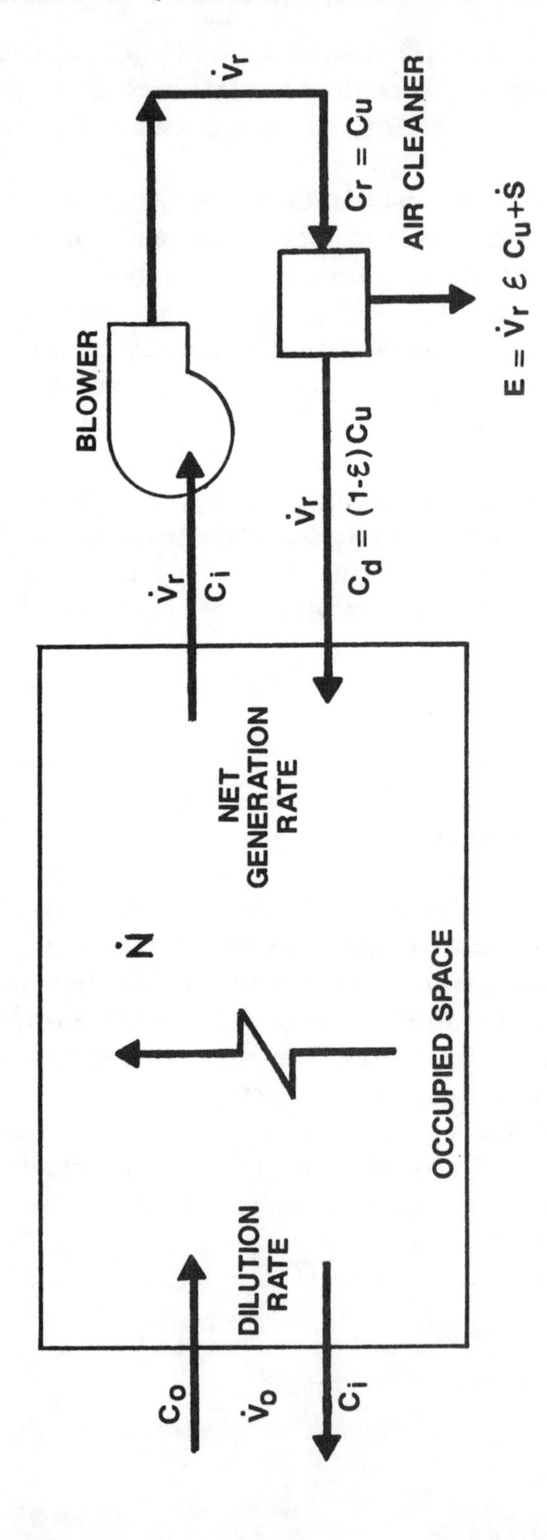

Figure 10-1. Single-Compartment, Uniformly-Mixed, Steady-State Model for Indoor Air Quality Control.

A mass balance of this uniformly-mixed space in steady state may be expressed by:

$$C = C_i - C_o = \frac{N - E}{V_o}$$

This model indicates that the indoor air concentration of a contaminant, C_i, will exceed the outdoor concentration, C_o, unless the removal rate exceeds the net generation rate. This control method is commonly used for clean rooms. This relationship also identifies the desired control strategy to achieve an acceptable ΔC to: (a) reduce the net operation rate, (b) apply techniques that will increase removal rate, and (c) increase the dilution rate.

10.2.2 Removal Rates

The removal rate of particulates and gaseous contaminants may be consisted of two components: passive removal mechanisms and active removal mechanisms. During passive removal, the concentrations of particulates and gaseous contaminants may be reduced by settling, condensation, or sorption. Reactive gases such as ozone, sulfur dioxide, and nitrogen dioxide may be adsorbed or react with building surfaces. However, procedures to calculate passive removal rates of gases and vapors are not well developed at this time. In many cases, the recycled air is cleaned and then returned to the building for energy conservation purposes. The active removal rate of a contaminant from recycled air may be expressed as:

$$E = V_r \epsilon C_u$$

where:

V_r = Volumetric airflow through air cleaner

ϵ = Air cleaner efficiency

C_u = Concentration of contaminant upstream of air cleaner.

10.2.3 Dilution Control

Dilution control is the most commonly used method of IAQ control in the introduction of outdoor air for dilution of indoor air concentrations. Dilution control may be the most energy-intensive and costly of the control methods available today. There are two different types of dilution controls: infiltration and natural ventilation, and mechanical ventilation.

Infiltration is usually considered to be undesirable air leakage through cracks, joints and connections in a building, whereas natural ventilation is usually considered to be desired air exchange through intentional openings in a building such as doors and windows. Because of the large number of variables associated with infiltration and natural ventilation, it is difficult to accurately estimate these variables. However, there are simple models to estimate both the infiltration and natural ventilation rates.

For the case of no recirculated air within an occupied space (100% outdoor air) and no removal control, the indoor concentration varies inversely with the dilution rate, V_o. This relationship is the basis for the prescriptive ventilation rates commonly specified in ventilation codes and standards. However, the assumption of uniform mixing within an occupied space may lead to significant error in correctly estimating the amount of dilution air required for a given location within an occupied space.

10.2.4 Ventilation Effectiveness/Efficiency

The single-compartment model, shown in Figure 10-1, assumes that the air within the occupied space is uniformly mixed. However, thermal and contaminant stratification can occur within the occupied space, resulting in occupant exposures much higher than predicted by models that ensure uniform mixing. The effectiveness of the ventilation system for IAQ control depends on two system characteristics: the room air exchange rate, and airflow patterns within the room.

If the room air distribution is not sufficient enough to dilute or remove indoor air contaminants from the location of most likely exposure, the effectiveness of the system will be impaired as excessive air exchange rates will probably be used to compensate, with expected results of increased energy consumption and nonuniform mixing. The air distribution patterns within the room may be as important to the effectiveness of the ventilation system as the room air exchange rate.

Mechanical ventilation systems in large office buildings are designed to satisfy space-conditioning loads and to maintain acceptable IAQ. ASHRAE Standard 62-1989 (10) specifies alternative procedures to achieve an acceptable IAQ. The IAQ procedure provides a direct solution by restricting the concentrations of contaminants to some specified, acceptable levels. It incorporates both the quantitative and subjective evaluations. This procedure was first introduced in ASHRAE Standard 62-1981 (11) to permit innovative, energy-conserving practices allowing the designer to use whatever amount of outdoor air needed if it could be shown that the concentrations of indoor air contaminants were held below the recommended levels.

The more common designs are based on the ventilation rate (VR) procedure also in ASHRAE Standard 62-1989. According to the VR procedure, IAQ is considered acceptable if the required rates of acceptable outdoor air are provided for the occupied spaces. These ventilation rates for various occupied spaces are listed in ASHRAE Standard 62-1989 (10). They were selected to reflect the consensus that providing acceptable outdoor air at these rates would achieve an acceptable IAQ by reasonably controlling carbon dioxide, particulates, odors, and other contaminants common to those spaces. The outside air requirements are for well-mixed conditions where ventilation effectiveness approaches 100%. The question that arises is how much air should be provided to compensate for a loss in ventilation effectiveness. Inadequate circulation/distribution of air can lead to some poorly ventilated occupied areas. The concept of ventilation effectiveness has been developed to quantify the air distribution characteristics of a ventilated space.

The volumetric flow rate of outdoor air supplied and the air exchange rate have been used as parameters for assessing the performance of general ventilation systems. The use of these parameters is based on the assumption that both air and the contaminants are uniformly distributed throughout the ventilated space (complete mixing). This assumption is almost never true, nor is complete mixing, in general, the most efficient principle of ventilation (12).

Quantifying the performance of ventilation systems raises the question of what terms to express the following factors (12):

- Distribution of the ventilated air in a room;
- Time it takes to replace (i.e., exchange of air present in a room);
- Factors that determine the detection, distribution, and removal of a contaminant released in a room; and
- Expressions for the efficiency of air exchange, dilution, and removal of contaminants.

10.2.4.1 Definitions and Basic Relations

Air Exchange Efficiency

One way of characterizing the ventilation process is to observe the frequency of air change in a ventilated space. Efficiency can then be determined by comparing the actual exchange frequency with the ideal one (13). The inverse of the frequency is a time constant, referred to as the turnover time, transit time, residence time or mean age, τ_n, and is defined by:

$$\tau_n = \frac{V}{\dot{V}}$$

where:

V = Room volume

$\dot{V}$ = Air flow.

These parameters are shown in Figure 10-2. The real and measurable transit time is system-invariant and independent of the airflow patterns in the room. It represents the shortest possible average residence time for the total air mass in the room (13).

If there is complete mixing in the room, the probability for a "new lump" of air entering the room to stay in, and hence replace an "old lump" of air, is the same as the probability of a "new lump" of air to leave the room without replacing any "old lump" of air. The result is that the average residence time for the room, τ_r, is twice the transit time for ventilation airflow, τ_n. This result can be shown using a statistical analysis (13). If the airflow is unidirectional (plug flow), the average residence time for the room air is exactly equal to the transit time for the ideal case. Tendency toward unidirectional flow is called displacement flow (13). Examples of ventilation flow patterns are shown in Figure 10-3.

The average air exchange efficiency is given by:

$$\eta_a = \frac{\tau_n}{\tau_r} \times 100\%$$

The values of η_a for various ventilation conditions are shown in Table 10-1.

Table 10-1. Average Air Exchange Efficiency.

Flow Types	Average Air Exchange Efficiency (η_a)
Stagnant flow (shortcircuiting)	$0 < \eta_a < 50\%$
Complete mixing	50%
Displacement flow	$50\% < \eta_a < 100\%$
Plug flow (unidirectional or piston flow)	100%

The above parameters can be determined using a tracer gas technique. The average η_a only reflects the gross flow patterns in a room. It is sometimes important to consider conditions in the occupied zone. At some points in the room, the mean age of the air may be different than the room average mean age. For example, if the air comes into the occupied zone quicker than other parts of the room, the mean age of the air is lower than the room average mean age. The ratio between the average and local age is defined as local air exchange indicator, ϵ_a (8), and is:

$$\epsilon_a = \frac{\tau_r}{2\tau_i} \times 100\%$$

where:

τ_r = Average residence time for the room (as defined before)
τ_i = Local mean age of air.

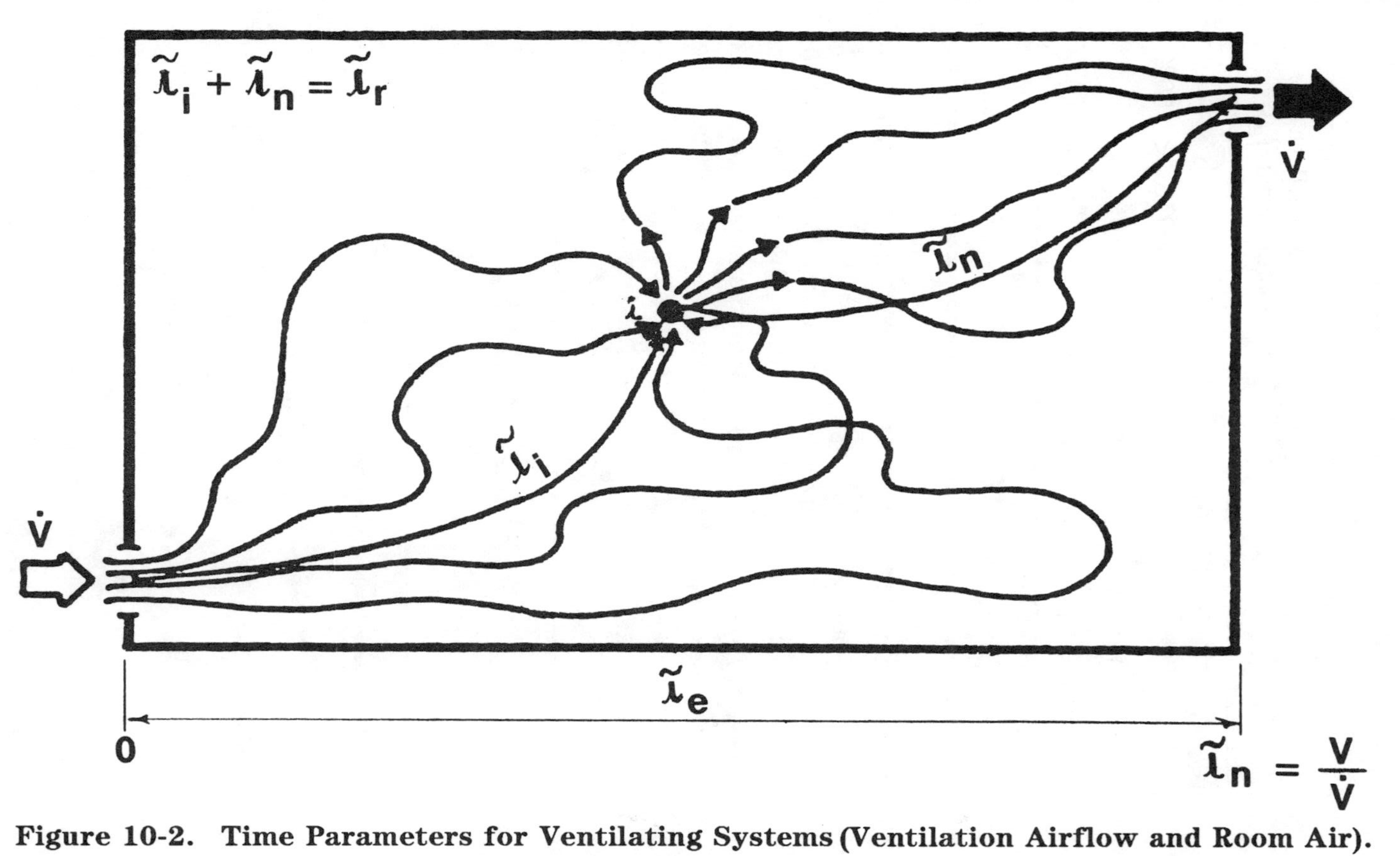

Figure 10-2. Time Parameters for Ventilating Systems (Ventilation Airflow and Room Air).

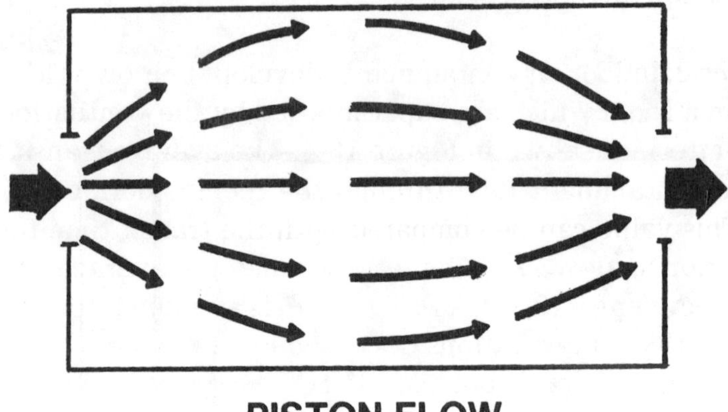

PISTON FLOW

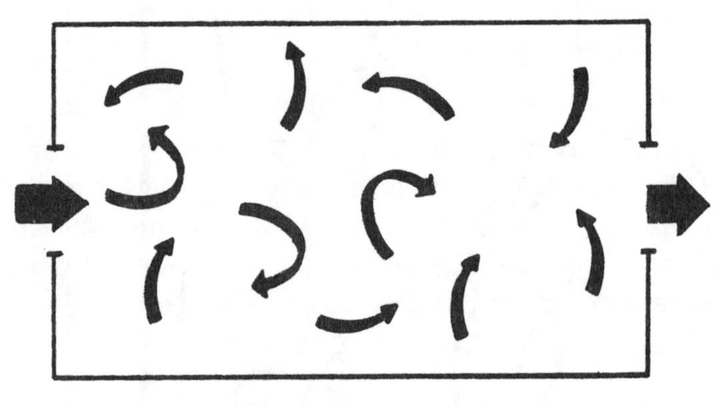

PERFECT MIXING

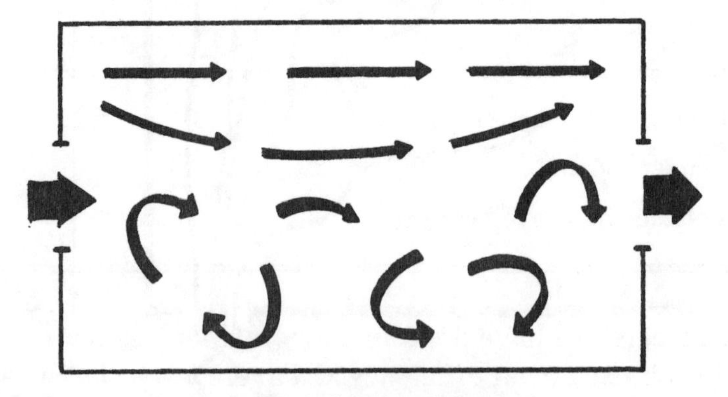

SHORT CIRCUITING

Figure 10-3. Examples of Ventilation Flow Patterns.

Ventilation Effectiveness

In general, indoor air contaminants develop their own flow patterns in a room which are superimposed by the ventilation airflow pattern as shown in Figure 10-4. The average transit time for the contaminant flow through the room is defined as τ^c_t (13). This value can be compared with the transit time for the ventilation airflow, τ_n. The ratio is called the average ventilation effectiveness (8, 13, 14). The ratio is also equal to the ratio between the concentrations of contaminants in the exhaust air and average concentrations of the contaminants in the room at steady state. It is:

$$\epsilon^c_v = \frac{C_e(\infty)}{C_i(\infty)} = \frac{\tau_n}{\tau^c_t}$$

Values for the average ventilation effectiveness are shown in Table 10-2 (13, 14). These values show that a design goal should be to achieve an average ventilation effectiveness greater than 1. This can be accomplished by keeping the transit time for the contaminant, τ^c_t, as short as possible. This is precisely the principle of a local exhaust system.

Table 10-2. Range of Average Ventilation Effectiveness.

Flow Type	Average Ventilation Effectiveness (ϵ)
Stagnant flow (shortcircuiting)	$0 < \epsilon^c_v$
Complete mixing	1
Contaminant source near exhaust	approaches ∞

Some researchers have not found any relationship between the average air exchanger efficiency and average ventilation effectiveness (15). However, a relationship developed by Skäret between the average ventilation effectiveness and contaminant removal efficiency, η^c_v, is:

$$\eta^c_v = \frac{\epsilon^c_v}{1 + \epsilon^c_v} \times 100\%$$

where:

$$\epsilon^c_v = \frac{\eta^c_v}{100 - \eta^c_v}$$

Two-Zone Models

In general, flow patterns in a room are quite complicated due to: (a) warm and cold surfaces, (b) unevenly distributed heat sources, (c) supply diffuser design, (d) air supply and exhaust locations, (e) infiltration and exfiltration, and (f) unevenly distributed pollutants with regard to thermal stratification. This complicated situation is modeled in practice using a two-zone model with reasonably good results (8, 13, 14, 16, 17, 18). It has been used extensively by scientists in Scandinavia for designing displacement ventilation systems (13). Simplified models have been used where there is no infiltration, exfiltration, or recirculation to calculate ventilation effectiveness for various supply and exhaust configurations as shown in Figure 10-5.

In Figure 10-5, Q is the airflow into and out of the total space volume, and BQ is the flow between the zones where B is a fraction which ranges from zero to infinity, corresponding to complete mixing between the zones. In each zone, the air is assumed to be perfectly mixed. Case A in Figure 10-5 is typical of most office buildings with the supply and exhaust located at the ceiling. It is suggested that this may lead to a situation in which a significant portion of the supply air shortcircuits the occupied zone. This model has been used extensively by Sandberg (12), Skäret (8, 13), and others.

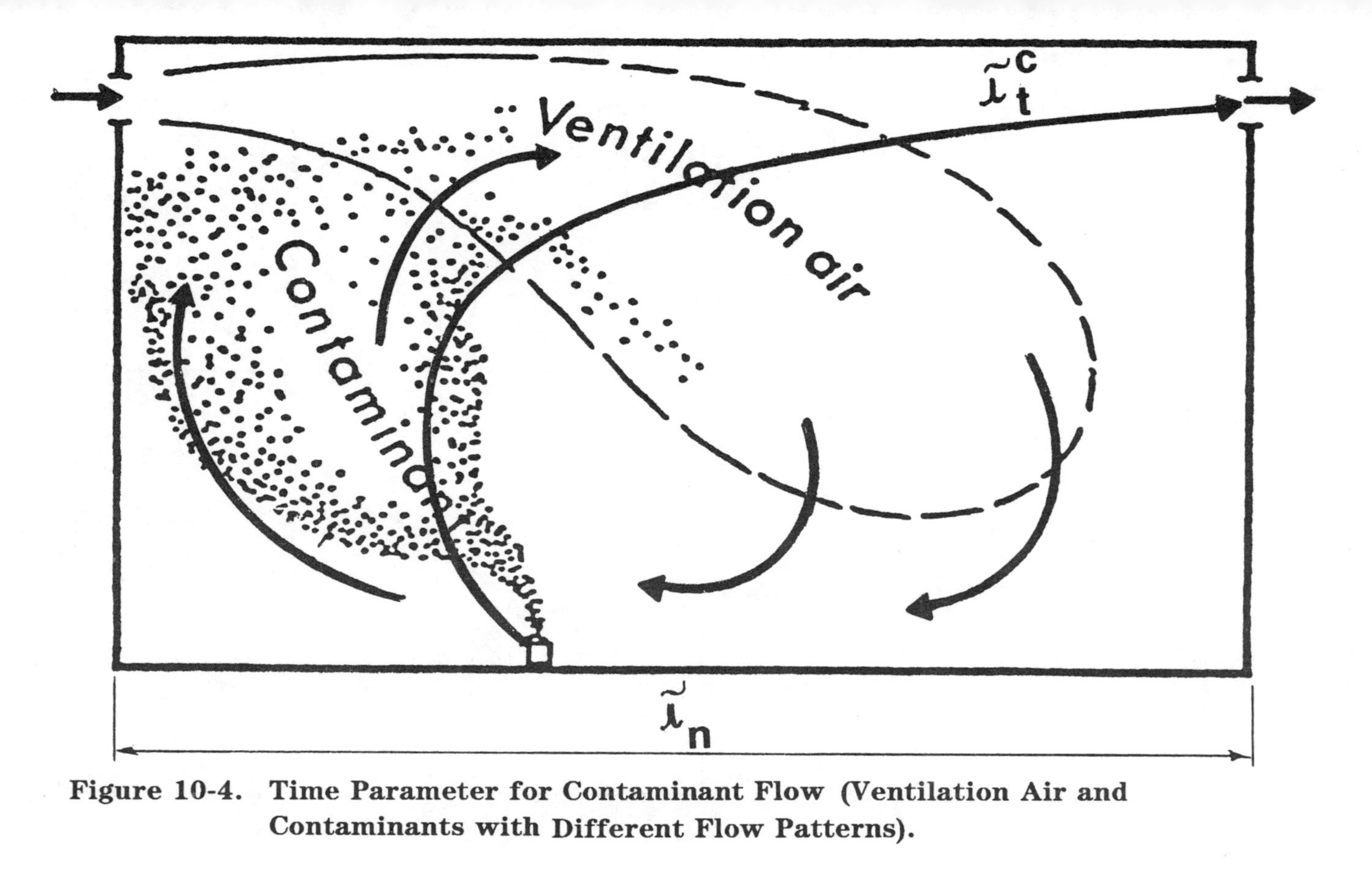

Figure 10-4. Time Parameter for Contaminant Flow (Ventilation Air and Contaminants with Different Flow Patterns).

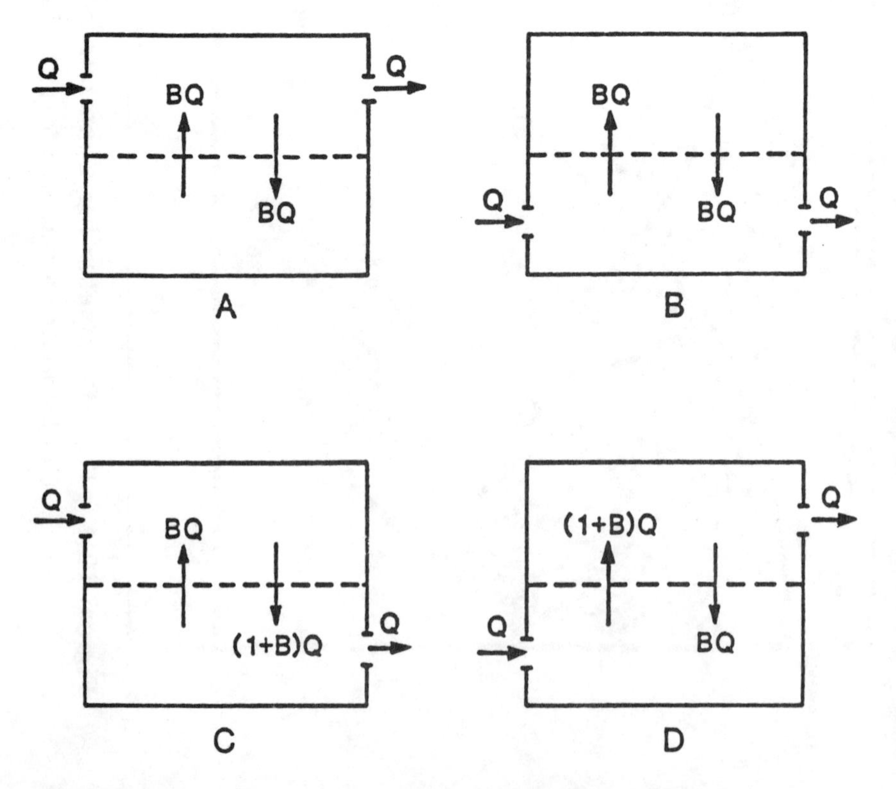

Figure 10-5. Flow Diagrams of Two-Zone Models.

For an actual installation, mechanical ventilation systems almost always employ recirculation of exhaust air as an energy-saving measure (18). A two-zone approach has been developed to include recirculation and is based on the office ventilation shown in Figure 7-1 in Chapter 7, Part 2. This is a stratification model presented in ASHRAE Standard 62-1989 (10). In Figure 7-1, the stratification factor S defines the fraction of the supply airflow which bypasses the occupied part of the room, and R is the fraction of the exhaust air which is recirculated and mixed with the new outside air. The quantity (1-S) is mixed with the room air (Refer to Figure 10-6).

In Figure 10-6, ventilation efficiency is defined in terms of the stratification factor S and the recirculation factor R, and can be thought of as the percent of outdoor air entering the room that ventilates the occupied zone (11, 16). The ventilation efficiency is defined (11, 16, 17) as:

$$E = \frac{1 - S}{1 - RS}$$

A similar definition was developed by Woods where S is referred to as the room supply air bypass factor. The calculation of this bypass factor is complicated and can be found in the literature (19).

To ensure acceptable IAQ, care must be taken in locating the supply and exhaust devices. For example, the common procedure of locating both the supply and exhaust devices at the ceiling has been shown to result in a ventilation efficiency of less than 50%. Typical ventilation efficiencies and stratification factors are shown in Figure 10-7 and Table 10-3 (16), respectively.

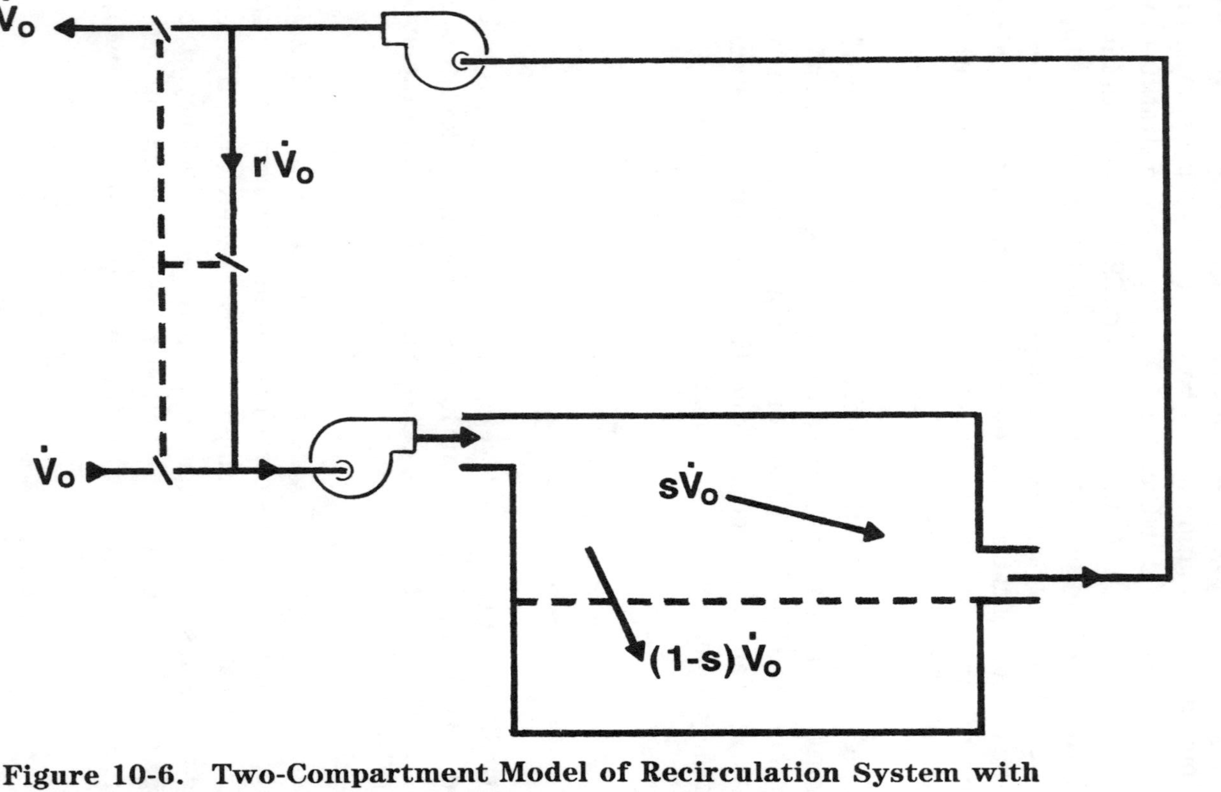

Figure 10-6. Two-Compartment Model of Recirculation System with Stratification to Predict Ventilation Efficiency.

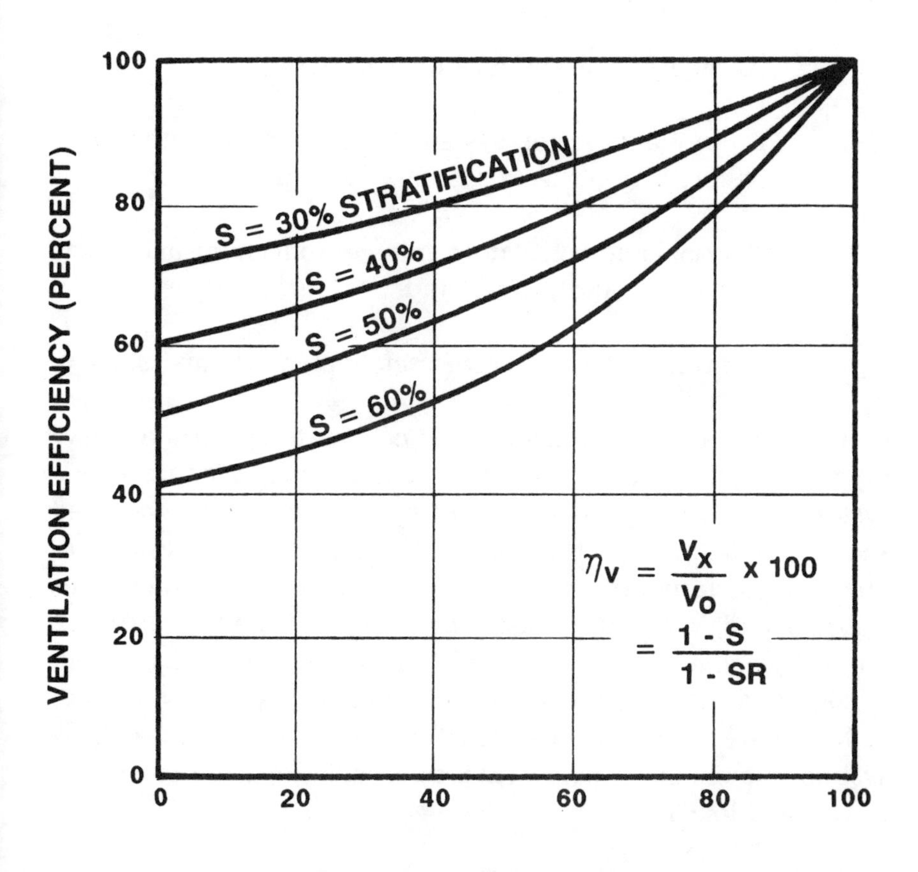

Figure 10-7. Ventilation Efficiency.

Table 10-3. Stratification Factors.

Load	Band Room 300	Chorus Room 162	Chorus Room 164	Main Return
Heating	0.57	0.47	0.39	0.38
No Load	0.43	0.58	0.56	0.44
Cooling	0.37	0.33	0.46	0.38

The basic components of heating, ventilating, and air-conditioning (HVAC) systems include:

- An energy supply; (gas or electricity)

- Energy conversion systems; (hot water generation or boiler, refrigeration system, humidifier, etc.)

- Thermal and ventilation transport mechanisms; (duct, ductwork, and piping; and

- Control systems; (thermostats, valves, dampers, etc.).

Central, forced-air systems may be constant-air-volume (CAV) or variable-air- volume (VAV). The basic difference is that a CAV system provides the same amount of airflow into an occupied space, independent of thermal load, whereas a VAV system reduces the airflow rate into the occupied space as a function of thermal load. VAV systems are inherently more energy-efficient where variable thermal loads exist, as the power reduction can be significant. Conversely, VAV systems may at times operate at less than the required ventilation capacities because these systems reduce their airflow rates to occupied spaces in response to thermal loads.

10.2.5 Air-Cleaning Systems

Meckler and Janssen (2) proposed the use of air cleaners to reduce outdoor air requirements. The objective of air cleaners is to increase the fraction of return air and thus reducing the energy required to condition the ventilation air. They found that there are a number of design considerations which affect the performance of air-cleaning systems. The type of air cleaners used, their location in the system, and the amount of air passed through the air cleaner all affect the amount of reduction in indoor air contaminants.

Particles are the easiest contaminants to remove from the air. The general types of particulate air cleaners include: (a) media filters, (b) electrostatic air cleaners, (c) adsorbers/absorbers, (d) centrifugal separators, and (e) air washers. Chemical reagents can be used to remove gases such as sulfur dioxide from power plant boiler gases. However, the high cost and the need for regular maintenance make air washer systems unattractive for use in commercial buildings.

The model developed by Meckler and Janssen (2) is shown in Figure 7-1 (Chapter 7, Part 2). The air cleaner can be placed in one of two different locations: recirculated stream or the mixed stream, identified as A and B, respectively in Figure 7-1. The volumetric flow through the air cleaner and its effectiveness determine the amount of material that can be removed. When the volumetric flow is also reduced in a VAV system, the contaminant removal capacity is reduced. This effect is incorporated in the flow reduction factor, F_r. The filter effectiveness is given by E_f. In general, the effectiveness increases as the velocity through the air cleaner is reduced. The recirculation flow factor, R, is the fraction of the return air that is recirculated. As the ventilation requirements increase, the amount of outdoor air needed to dilute the occupied space increases. This causes the recirculation factor to decrease. The supply airflow can be either constant- or variable-volume, and the supply air temperature can be either constant or variable. If the supply air temperature is constant, the supply air volume must be varied to follow the thermal load.

10.2.6 Multi-Compartment Models

Sparks et al. (3) developed a multi-compartment model based on a well-mixed mixing model. They noted that air movement in a building consists of: (a) natural air movement between rooms, (b) air movement driven by a forced-air system, and (c) air movement between the building and the outside.

The pollution concentration in a room is calculated by a mass balance of various pollutant flows. The single-room mass flows are shown in Figure 10-8. Sparks et al. (3) extended the analysis to multiple rooms by devising a system of differential equations. The amount of air entering a room from all sources (HVAC, outdoors, other rooms, etc.) must equal the amount of air leaving the room. They also assumed a well-mixed model. Each differential equation represents a mass balance for each room. The resulting differential equations can be solved using standard numerical methods, resulting in concentration profiles (with respect to time) for each room.

The model requires the user to enter the inter-room airflows. Calculation of the inter-room airflows requires knowledge of the temperatures and pressure gradients between rooms. This information is generally not available. Sparks et al. (3) found that inter-room airflows do not have to be estimated with high precision to produce good results.

10.2.7 Goodfellow Consultants Inc. Model

The numerical model developed by Goodfellow Consultants Inc. (GCI) is an unsteady, multi-cell model which includes bulk flows between each cell, infiltration/exfiltration to each cell, recycled air to each cell, time-dependent emission sources from each cell, and mixing efficiency of each cell.

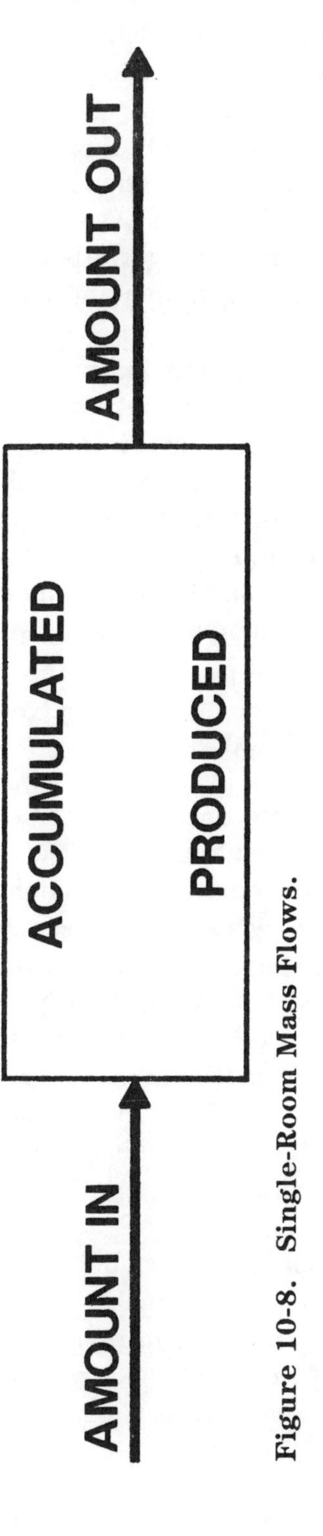

Figure 10-8. Single-Room Mass Flows.

The model has a menu-driven, data-input user interface for easy use. It can calculate concentration profiles for a maximum of 15 cells. The input parameters include the number of cells (rooms), cell volume, initial cell concentration, infiltration flowrate and concentration, exfiltration flowrate, HVAC flow rates and interconnecting flow rates between cells, and the emission rates. The block diagram in Figure 10-9 shows step-by-step procedures for calculating contaminant concentrations.

There is a wide range of emission rates available in the model. The emission rates defined in the numerical model are: (a) a constant source in the entire time interval, (b) a constant source for a set time interval, (c) an exponential decay from an initial value, (d) an exponential increase from an initial value, (e) an asymptotic increase to a final value, (f) a linear increase/decrease from an initial value, and (g) a combination of these types of emission rates. Several emission rates, N, are presented in Figure 10-10.

10.2.8 Generation Rates

The most significant factor to be quantified in IAQ models is the generation rate of the contaminants. There are several thousands of gases, vapors, particulates, and radionuclides that are emitted from three primary sources: (a) occupants of a building, (b) materials and furnishings within an occupied space, and (c) processes conducted within an occupied space. The emission sources vary from a ''single point'' (occupants, tobacco smoke, etc.) to large ''surface areas'' (building materials, out-gassing, etc.). Contaminant emissions may be independent of other indoor environmental conditions (e.g., tobacco smoke) or may be interactive with the environment (e.g., bacterial growth).

The simplest assumption for modeling emissions is to set the rates independent of other environmental factors. With this assumption, the contaminant emission rates may be considered to be constant for specific time intervals which are dependent on the mobility of the source (e.g., occupants) or by the frequency of

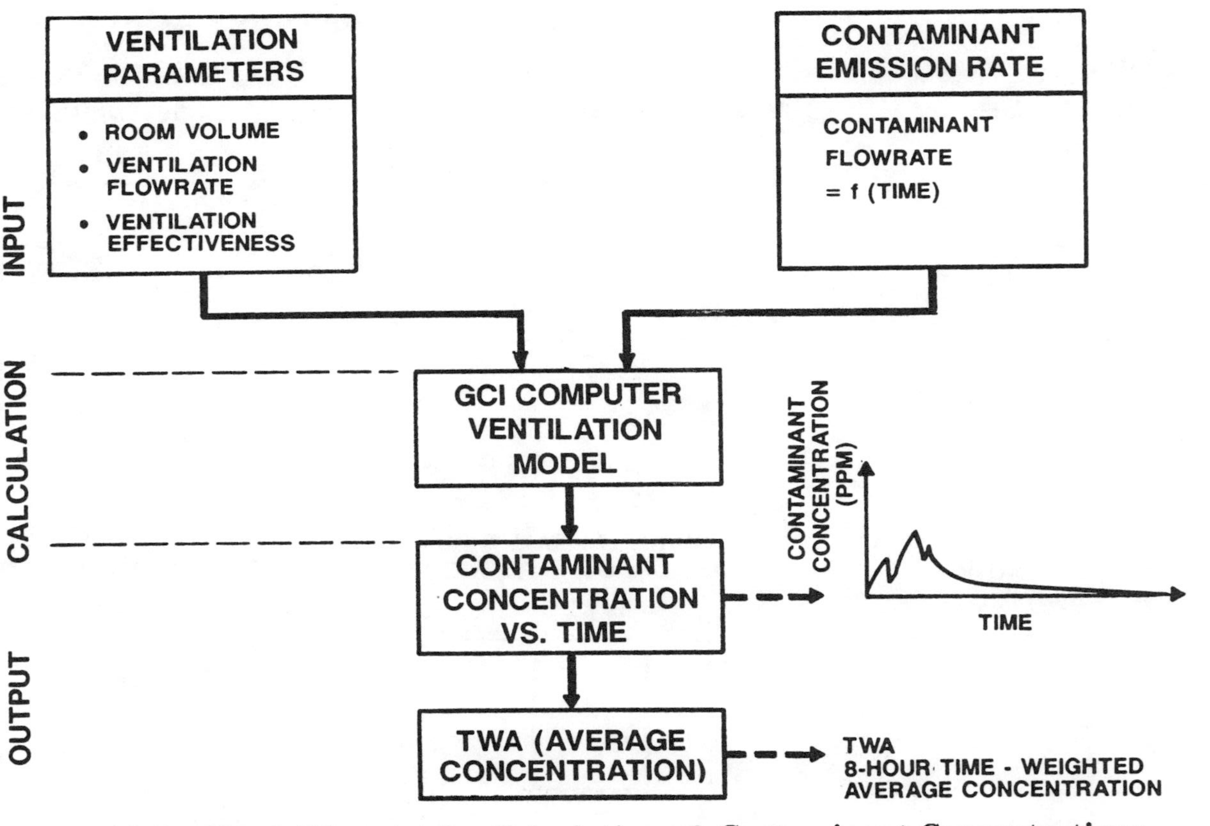

Figure 10-9. Block Diagram for Calculation of Contaminant Concentrations.

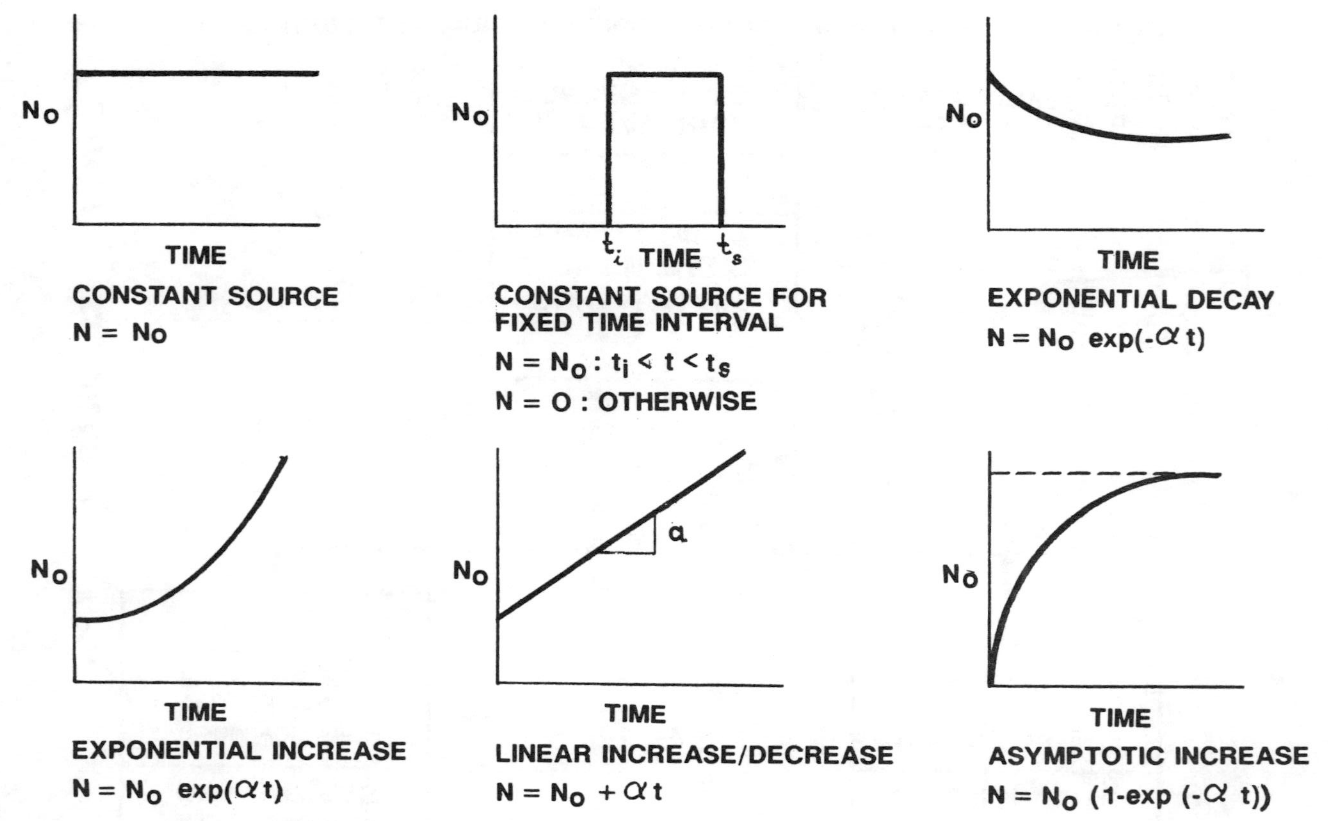

Figure 10-10. Emission Rates in GCI Model.

the occurrence. Although the assumption of emissions that are independent of environmental factors simplifies the calculations, significant errors in predicting exposures can be introduced.

There are two types of interactions: interactions with thermal factors and interactions with other contaminants. Formaldehyde is an example of a contaminant affected by thermal interactions. Net generation rates may double if the temperature increases by 6 °C. Radon is an example of a contaminant affected by interactions with other contaminants. When radon is in equilibrium with its progeny, the radiation dose may be 500 times that of radon itself.

In this model, it is important to accurately estimate the emission rate. The emission rate is highly dependent of the air velocity over the surface where the emission occurs. If the airflow is zero, the emission rate is primarily through stagnant diffusion. Conversely, if the airflow is non-zero, the emission rate increases markedly. The GCI model can calculate the concentration profile for systems where the emission rate is a function of time. The effects of air change rate and emission rate on contaminant concentration are shown in Figure 10-11.

In Figure 10-12, the emission rate (g/s) varies considerably as a function of time. In the first 100s, the emission rate is constant (5 g/s). Between 100s and 200s, the emission rate increases linearly at a rate of 0.1 g/s. From 200s to 300s, the emission rate decreases exponentially at a rate of exp(-0.01t) to 5.5 g/s. Between 300s and 350s, the emission rate decreases linearly at a rate of 0.05 g/s. From 350s to 400s, the emission rate is constant and equal to 8 g/s. After 400s, the emission rate is zero.

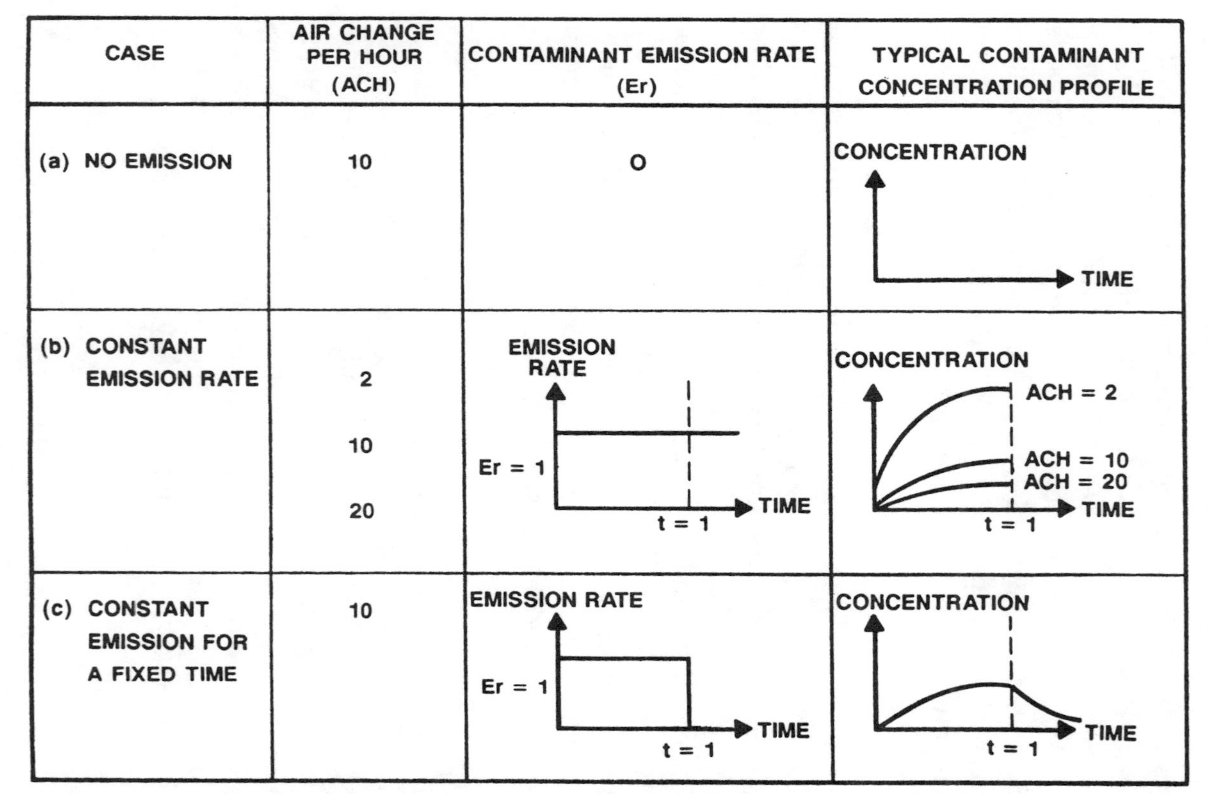

Figure 10-11. The Effects of Air Change Rate and Emission Rate on Contaminant Concentration.

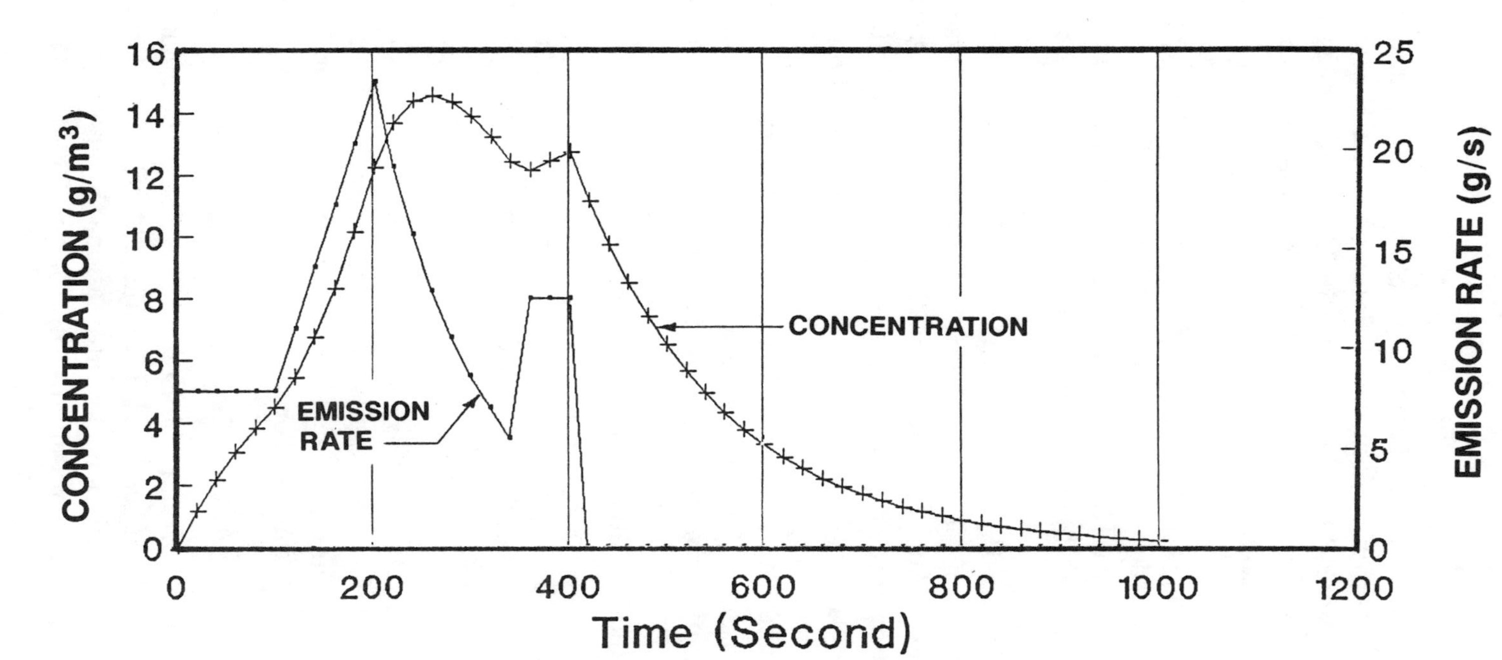

Figure 10-12. Multiple Emission Rates (Single-Cell Model).

References

1. Horstman, R.H. "Predicting Velocity and Contamination Distribution in Ventilated Volumes Using Navier-Stokes Equations." Proceedings of ASHRAE Conference IAQ '88: Engineering Solutions to Indoor Air Problems, Atlanta, GA, pp. 209-230, 1988.

2. Meckler, M. and Janssen, J.E. "Use of Air Cleaners to Reduce Outdoor Air Requirements." Proceedings of ASHRAE Conference IAQ '88: Engineering Solutions to Indoor Air Quality Problems, Atlanta, GA, pp. 130-147, 1988.

3. Sparks, L.E.; Jackson, M.D.; and Tichenor, B.A. "Comparison of EPA Test House Data with Predictions of an Indoor Air Quality Model." Proceedings of ASHRAE Conference IAQ '88: Engineering Solutions to Indoor Air Quality Problems, Atlanta, GA, pp. 251-264, 1988.

4. Crommelin, R.D. and Buringh, E. "Validation of a Multiple Cell Model for the Prediction of Air Temperatures and Pollution Concentrations by Measurements in an Industrial Hall." TNO Division of Technology for Society, Department of Indoor Environment, Netherlands, 1988.

5. Awbi, H.B. "Numerical Solution of Air Movement in Rooms." Ventilation '88 Symposium, London, England, Sept., 1988.

6. Siurna, D.L. and Bragg, G.M. "Stochastic Modeling of Room Air Diffusion." Ventilation '85 Symposium, Proceedings of the 1st International Symposium on Ventilation for Contaminant Control, Toronto, Ontario, Canada, pp. 121-135, Oct., 1985.

7. Fletcher, B. and Johnson, A.E. "The Accumulation of Gases in Ventilated and Unventilated Enclosures." Proceedings of Ventilation '85: International Symposium on Ventilation for Contaminant Control, pp. 333-354, 1985.

8. Skäret, E. "Industrial Ventilation Model Tests and General Development in Norway and Scandinavia." Ventilation '85, Oct., 1985, Toronto, Canada, Elsevier Science Publishers, Amsterdam, pp. 19-32, 1986.

9. Dellagi, F.; Dumaine, J.Y.; and Aubertin, G. "Numerical Simulation of Air Flows - Applications to the Ventilation of a Paint-Booth." Ventilation '85, Oct., 1985, Toronto, Canada, Elsevier Science Publishers, Amsterdam, pp. 391-403, 1986.

10. ASHRAE Standard 62-1989. "Ventilation for Acceptable Indoor Air Quality." Atlanta, GA, 1989.

11. ASHRAE Standard 62-1981. "Ventilation for Acceptable Indoor Air Quality." Atlanta, GA, 1981.

12. Sandberg, M; Blomquist, C; and Sjoberg, M. "Efficiency of General Ventilation Systems in Residential and Office Buildings - Concepts and Measurements." Ventilation '85, Oct., 1985, Toronto, Canada, Elsevier Science Publishers, Amsterdam, pp. 323-332, 1986.

13. Skäret, E. "Ventilation by Displacement - Characterization and Design Implications." Ventilation '85, Oct., 1985, Toronto, Canada, Elsevier Science Publishers, Amsterdam, pp. 827-841, 1985.

14. Rodahl, E. "Ventilation Effectiveness - Past and Future." Proceedings of the 4th International Conference on Indoor Air Quality and Climate, West Berlin, Vol. 4, pp. 57-68, 1987.

15. Antti, M; Tapio, H.; and Olli, S. "Air Quality and Ventilation Efficiency in Residential and Office Buildings." Proceedings of the 4th International Conference on Indoor Air Quality and Climate, West Berlin, Vol. 3, pp. 301-306, 1987.

16. Janssen, J.; Hill, T.J.; Woods, J.E.; and Maldonado, E.A.B. "Ventilation for Control of Indoor Air Quality: A Case Study." *Environ. International*, 8, pp. 487-496, 1982.

17. Woods, J.R. "Status-Ventilation Models for Indoor Air Quality." Ventilation '85, Toronto, Canada, Oct., 1985, Elsevier Science Publishers, pp. 33-52, Amsterdam, 1986.

18. Persily, A.K. "Ventilation Effectiveness in Mechanically Ventilated Office Buildings." U.S. Department of Commerce: National Bureau of Standards (NBSIR 85-3208), 1985.

19. Rask, D.R.; Woods, T.E.; and Sun, J. "Ventilation Efficiency." Building Systems: Room Air and Air Contaminant Distribution Symposium, Dec., 1988.

11. ZONING FOR INDOOR AIR QUALITY: AIR DISTRIBUTION AND DESIGN TECHNIQUES

Milton Meckler, P.E.
President, The Meckler Group
Encino, California

Page

11.1 Introduction

Recently, heating, ventilating, and air-conditioning (HVAC) designers have been held more accountable for the health and safety of the occupants of the buildings. This is, perhaps, a direct result of an increased public awareness of the health and safety risks associated with some uncontrolled contaminants and a growing awareness in general public that their health may be at risk. Although energy efficiency is a necessary criterion for good engineering design, it must not be allowed to overshadow valid environmental concerns for occupant well-being and protection against undue discomfort, sickness, and loss of productive time.

HVAC systems must function to provide an environment where the temperature, air distribution, humidity, and indoor air quality (IAQ) are maintained within the prescribed operating ranges.

A general consensus already exists within the American Society of Heating, Refrigerating, and Air-Conditioning Engineers (ASHRAE) with regard to the first three parameters. Unless a case can be made for air quality zoning, the resulting air distribution design approach generally becomes a function of temperature and/or humidity. ASHRAE Standard 62-1989 provides criteria for the application of the IAQ procedure as an alternative to the more prescriptive ventilation rate (VR) procedure. Table 11-1, used in conjunction with the IAQ procedure defines seven classes of air distribution systems which relate filter locations in terms of space concentration of contaminants and required outdoor and recirculation air rates.

Most documented cases of sick-building syndrome (SBS) or building-related illness (BRI) reflect a failure to diagnose or relate occupant discomfort or illness with building operations. SBS and BRI generally refer to a class of illnesses with potential health problems where construction materials and often maintenance chemicals combine with poor ventilation and air circulation and, result in inadequate treated outdoor and recirculated air supplied to the occupants of the building. Symptoms range from headache, nausea, eye strain, dizziness, rashes, persistent cough, dry throat, and some respiratory problems. Sources of SBS and BRI include formaldehyde, plasticizers, paint, wall and ceiling materials, cleaning agents, fiberglass and rock wool, carbon monoxide (CO), carbon dioxide (CO_2), adhesives, caulking, pesticides, various viable and nonviable microorganisms, and volatile organic compounds (1,2). For the psychological effects of sick buildings and preventive measures, refer back to Chapter 6, Part 1.

IAQ is directly related to outdoor air quality because ventilation employing outdoor air is essential to the replacement of oxygen consumed and to dilute indoor air pollutants to an acceptable level. For example, CO_2 exhaled indoors must be diluted to achieve desirable comfort and reduce odors. As the ventilation load increases, the amount of outdoor air needed to dilute the occupant-generated CO_2 increases.

Class	Required Recirculation Rate				Required Outdoor Air	Space Contaminant Concentration	Required Recirculation Rate
	Filter Location	Flow	Temperature	Outdoor Air			
I	None	VAV	Constant	100%	$V_o = \frac{N}{E_v F_r (C_s - C_o)}$	$C_s = C_o + \frac{N}{E_v F_r V_o}$	Not applicable
II	A	Constant	Variable	Constant	$V_o = \frac{N - E_v RV_r E_f C_s}{E_v (C_s - C_o)}$	$C_s = \frac{N + E_v V_o C_o}{E_v (V_o + RV_r E_f)}$	$RV_r = \frac{N + E_v V_o (C_o - C_s)}{E_v E_f C_s}$
III	A	VAV	Constant	Constant	$V_o = \frac{N - E_v F_r RV_r E_f C_s}{E_v (C_s - C_o)}$	$C_s = \frac{N + E_v V_o C_o}{E_v (V_o + F_r RV_r E_f)}$	$RV_r = \frac{N + E_v V_o (C_o - C_s)}{E_v F_r E_f C_s}$
IV	A	VAV	Constant	Proportional	$V_o = \frac{N - E_v F_r RV_r E_f C_s}{E_v F_r (C_s - C_o)}$	$C_s = \frac{N + E_v F_r V_o C_o}{F_r E_v (V_o + RV_r E_f)}$	$RV_r = \frac{N + E_v F_r V_o (C_o - C_s)}{E_v F_r E_f C_s}$
V	B	Constant	Variable	Constant	$V_o = \frac{N - E_v RV_r E_f C_s}{E_v [C_s - (1 - E_f) C_o]}$	$C_s = \frac{N + E_v V_o (1 - E_f) C_o}{E_v (V_o + RV_r E_f)}$	$RV_r = \frac{N + E_v V_o [(1 - E_f) C_o - C_s]}{E_v E_f C_s}$
VI	B	VAV	Constant	Constant	$V_o = \frac{N - E_v F_r RV_r E_f C_s}{E_v [C_s - (1 - E_f) C_o]}$	$C_s = \frac{N + E_v V_o (1 - E_f) C_o}{E_v (V_o + F_r RV_r E_f)}$	$RV_r = \frac{N + E_v V_o [(1 - E_f) C_o - C_s]}{E_v F_r E_f C_s}$
VII	B	VAV	Constant	Proportional	$V_o = \frac{N - E_v F_r RV_r E_f C_s}{E_v F_r [C_s - (1 - E_f)(C_o)]}$	$C_s = \frac{N + E_v F_r V_o (1 - E_f) C_o}{E_v F_r (V_o + RV_r E_f)}$	$RV_r = \frac{N + E_v F_r V_o [(1 - E_f) C_o - C_s]}{E_v F_r E_f C_s}$

Table 11-1. Effect of Filter Location and System Parameters on Required Outdoor Air, Space Contaminant Concentration or Recirculation Rate.

ASHRAE Standard 62-1989 recommends 15 cfm of outdoor air per person as the minimum needed to control occupant odors and guarantee that the concentration of CO_2 will not exceed 1000 ppm. It also recommends that the indoor air as a minimum not exceed the limits currently established by the Environmental Protection Agency (EPA) for outdoor air and the National Primary Ambient Air Quality Standards (NPAAQS). The EPA has mandated that the concentration of particulates should not exceed 50 $\mu g/m^3$ and 150 $\mu g/m^3$, for long and short terms, respectively (refer to Table 11-2). The Air Quality Standards Compliance Report, published monthly by the South Coast Air Quality Management District (SCAQMD) in southern California, has compared outdoor air quality at several designated air-monitoring stations in the South Coast Air Basin and Southeast Desert Air Basin to the state and federal ambient air quality standards. In this report, compiled January 1988 - June 1988, the concentration of particulates alone recorded in a 24-hour period in the outdoor air has exceeded both the state and federal standards in 1987 (3). For example, Figure 11-1 shows the particulate concentrations in the respirable range (PM 10) for February 1988. These concentrations have exceeded both the state and federal standards by far in comparison to others. Therefore, the use of outdoor air in Los Angeles as a dilution strategy for particulates has become highly questionable.

Since ASHRAE Standard 62-1989 requires indoor air not to exceed the NPAAQS in designing HVAC systems within the SCAQMD, it may be prudent to consider measures that will reduce the introduction of particulates that are potentially harmful to human health in the respirable range (4,5). The use of air recirculation with proper filtration and air-cleaning may provide a more viable alternative for particulate control in buildings located in the SCAQMD and in reducing the associated energy costs.

The IAQ procedure is an integral part of ASHRAE Standard 62-1989 and can be used to compute the amount of outdoor air required for known amounts of indoor air contaminants. The equations are presented in Table 11-1 for the more commonly used air distribution systems. In Table 11-1, one needs to spec-

AIR POLLUTANT	LONG TERM			SHORT TERM		
	CONCENTRATION		AVERAGE	CONCENTRATION		AVERAGE
	$\mu g/m^3$	ppm		$\mu g/m^3$	ppm	
SULFUR DIOXIDE	80.0	0.030	1 YEAR	365 (a)	0.14 (a)	24 HOURS
PARTICLES (PM 10)	50.0 (b)	—	1 YEAR	150 (a)	—	24 HOURS
CARBON MONOXIDE	—	—	—	40,000	35.00	1 HOUR
CARBON MONOXIDE	—	—	—	10,000 (a)	9.00 (a)	8 HOURS
OXIDANTS (OZONE)	—	—	—	235 (c)	0.12 (c)	1 HOUR
NITROGEN DIOXIDE	100.0	0.055	1 YEAR	—	—	—
LEAD	1.5	—	3 MONTHS (d)	—	—	—

(a) MAY BE EXCEEDED ONLY ONCE PER YEAR.

(b) ARITHMETIC MEAN.

(c) STANDARD IS ATTAINED WHEN EXPECTED NUMBER OF DAYS PER CALENDAR YEAR WITH MAXIMUM HOURLY AVERAGE CONCENTRATIONS ABOVE 0.12 PPM (235 $\mu g/m^3$) IS EQUAL TO OR LESS THAN 1.

(d) 3-MONTH PERIOD IS A CALENDAR QUARTER.

REFERENCE : ASHRAE Standard 62-1989

Table 11-2. National Primary Ambient Air Quality Standards Set by Environmental Protection Agency for Outdoor Air.

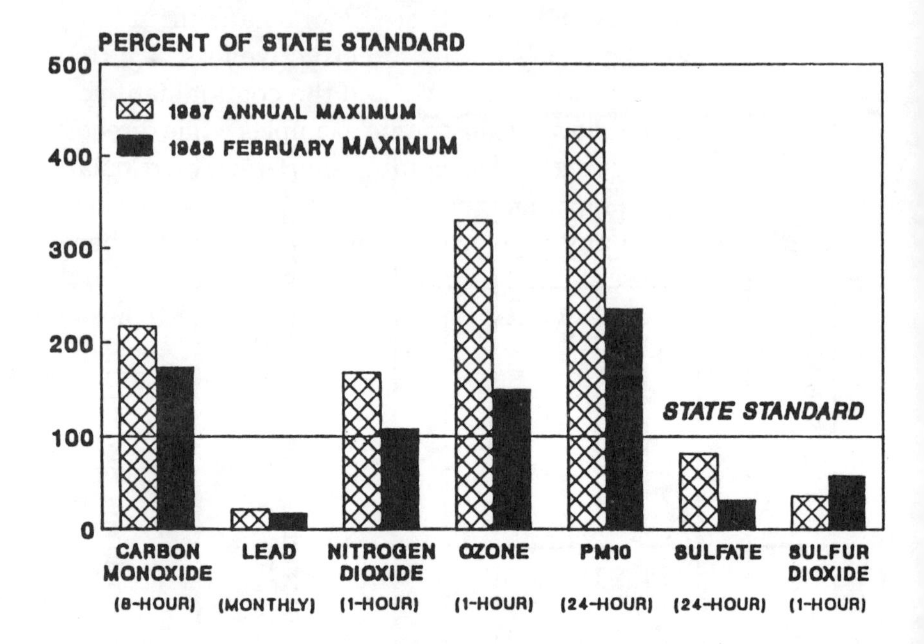

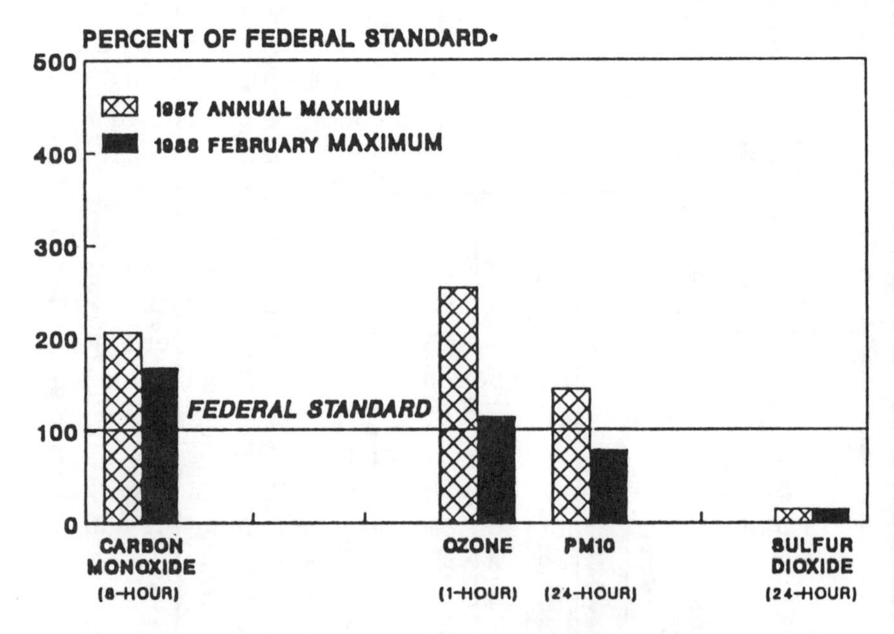

REFERENCE : Air Quality Standards Compliance Report Published by SCAQM DISTRICT.

Figure 11-1. Concentrations of Air Contaminants in February 1988 with Respect to State and Federal Standards.

ify the indoor contaminant level, C_s; outdoor contaminant level, C_o; filter effectiveness, E_f; ventilation effectiveness, E_v; flow reduction factor, F_r; return rate, V_r; and the contaminant generation rate, $\dot{N}$. The following illustrative examples while directed mainly toward satisfying the acceptable short-term particulate concentrations in compliance with NPAAQS, can just as easily be used to examine any other contaminant of interest. Therefore, there is a present need to develop a consistent methodology that can be used to evaluate how best to zone for IAQ (health needs) as well as temperature and humidity (comfort needs). These examples illustrate how the IAQ zoning procedure can be implemented.

11.2 Illustrative Examples Based on ASHRAE Standard 62-1989

The following examples illustrate the computer-assisted (6) use of IAQ procedure using the equations presented in Table 11-1 for variable-air-volume (VAV) systems at 100% full- and 50% part-load conditions. The 50% part-load condition is assumed to be a minimum to provide acceptable indoor contaminant levels. Also notice that the number of people in a zone and $\dot{N}$ are reduced to realistically reflect the reduced activities at 50% part-load conditions. Equations presented in Table 11-1 are solved directly for C_s and V_o and, simultaneously verified all flow balances (to and from the conditioned spaces). The building under study is a nominal 22,000-ft^2 office building in Burbank, California with dimensions shown in Figure 11-2. It contains four separate occupancy zones: (a) telecommunications, (b) conference rooms, (c) offices, and (d) shops. The calculated load summary in Table 11-3 shows square footage of each zone, individual zone loads, number of people present in each zone, and the volumetric flow rate of air supply in cfm.

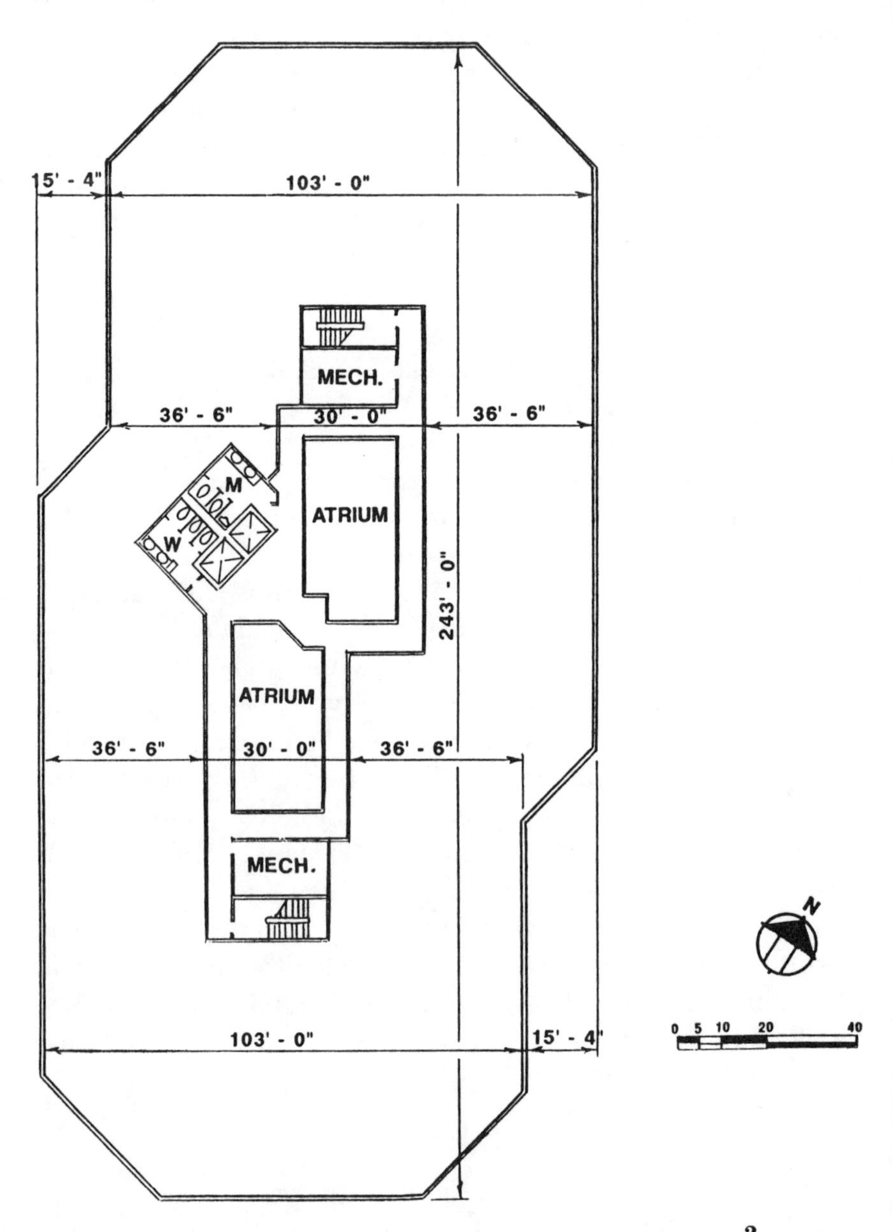

Figure 11-2. Floor Plan for a Nominal 22,000-FT^2 Office Building in Burbank, CA.

ZONE	ZONE AREA (FT^2)	ROOM SEN. LOAD (BTU/HR)	* NO. OF PEOPLE	ROOM LATENT LOAD (BTU/HR)	ROOM TOTAL LOAD (BTU/HR)	SUPPLY AIR (CFM)
TELECOMMUNICATIONS	5,000	280	300	57	337	12,740
CONFERENCE ROOMS	4,000	164	200	50	214	7,600
OFFICES	9,000	281	63	12	293	13,000
SHOPS	4,000	147	100	25	172	6,800

* BASED ON ASHRAE STANDARD 62-1989

Table 11-3. Calculated Load Summary and Volumetric Air Flow Rates in Each Zone of Office Building in Burbank, CA.

The following are four illustrative examples for various VAV systems. Notice that in the first three cases, the outdoor air is pretreated to $C_o = 5\ \mu g/m^3$.

11.2.1 Case 1. Conventional VAV System (Class VI in Table 11-1)

The following data are for the telecommunications zone.

At 100% full-load condition, the system parameters are:

Volume of space = 1133 m^3 (40,000 ft^3)

Number of people = 300

$\dot{N} = 15\ \mu g/min \cdot m^3$

$E_v = 0.8$

$V_s = 361\ m^3/min$ (12,740 cfm)

$E_f = 0.5$

$F_r = 1$

$C_o = 5\ \mu g/m^3$

At 50% part-load condition, the system parameters are:

Volume of space = 1133 m^3 (40,000 ft^3)

Number of people = 210

$\dot{N} = 10.5\ \mu g/min \cdot m^3$

$E_v = 0.8$

$V_s = 361\ m^3/min$ (12,740 cfm)

$E_f = 0.5$

$F_r = 0.5$

$C_o = 5\ \mu g/m^3$

To satisfy $C_s < 50\ \mu g/m^3$ and $C_s < 150\ \mu g/m^3$ at long- and short-terms, respectively, and a minimum outside supply air of 15 cfm/person at part- as well as full-loads:

$V_o = 4358$ cfm with $C_s = 89\ \mu g/m^3$ at 100% full-load,

$C_s = 110\ \mu g/m^3$ at 50% part-load, and

$OSA_{calc.} = 15$ cfm/person (100% full-load) and 21 cfm/person (50% part-load).

11.2.2 Case 2. Conventional VAV System (Class VII in Table 11.1)

The following data are for the telecommunications zone.

Volume of space = 1133 m^3 (40,000 ft^3)

Number of people = 300

$\dot{N} = 15\ \mu g/min \cdot m^3$

$E_v = 0.8$

$V_s = 361\ m^3/min$ (12,740 cfm)

$E_f = 0.5$

$F_r = 1$

$C_o = 5\ \mu g/m^3$

At 50% part-load condition, the system parameters are:

Volume of space = 1133 m^3 (40,000 ft^3)

Number of people = 210

$\dot{N}$ = 10.5 $\mu g/min \cdot m^3$

E_V = 0.8

V_S = 361 m^3/min (12,740 cfm)

E_f = 0.5

F_r = 0.5

C_o = 5 $\mu g/m^3$

To satisfy $C_S < 50\ \mu g/m^3$ and $C_S < 150\ \mu g/m^3$ at long- and short-terms, respectively, and a minimum outside supply air of 15 cfm/person at part- as well as full-loads:

V_o = 6115 cfm with C_S = 81 $\mu g/m^3$ at 100% full-load,

V_o = 3058 cfm with C_S = 137 $\mu g/m^3$ at 50% part-load, and

$OSA_{calc.}$ = 20 cfm/person (100% full-load) and 15 cfm/person (50% part-load).

11.2.3 Case 3. Modified Water Source Heat Pump/Packaged Terminal Air Conditioner System (Class VI in Table 11-1)

This system (7), a natural extension of the conventional water source heat pump (WSHP) system for cooling-load-dominated climates, requires less total installed refrigeration capacity and also consumes less energy than a comparable WSHP system (8). Peak electrical demand limiting is obtained through shifting on-peak electrical energy consumption to lower cost, and daily off-peak and mid-peak rates are available to customers (7, 9, 10).

Some indoor contaminants can be eliminated by energy-efficient filtering of return air, thus eliminating the need for excessive amounts of outdoor air. One of the methods of compensating for the effect of reduced ventilation rates in VAV systems involves the introduction of additional filtration and recirculation of ventilated air. This provides enhanced overall filtration to compensate for the reduced supply air in response to the diminished net space demands for heating and cooling. To compensate for the reduced supply air at part-loads and the resulting increased contaminant concentrations, an additional filter can be placed in the bypass duct for the individual WSHP or packaged terminal air conditioner (PTAC) and a VAV unit. This additional filter E-2 is shown in Figure 11-3.

Based on several IAQ assessment studies (4, 5), use of such a recycle filtration system can significantly increase the effectiveness of VAV distribution systems to control indoor air contaminant concentrations. Installing filters in both the return and bypass positions as shown in Figure 11-3, of a unitary WSHP or PTAC unit incorporating a bypass control system for adapting constant-volume air supply unitary units to provide a comparable VAV air distribution system to the conditioned space, also makes cleaner air automatically available at all part-load conditions (9, 10).

At 50% part-load condition, the system parameters for the telecommunications zone are:

Volume of space = 1133 m^3 (40,000 ft^3)

Number of people = 210

$\dot{N} = 10.5\ \mu g/min \cdot m^3$

$E_v = 0.8$

$V_s = 361\ m^3/min$ (12,740 cfm)

$E_f = 0.67$

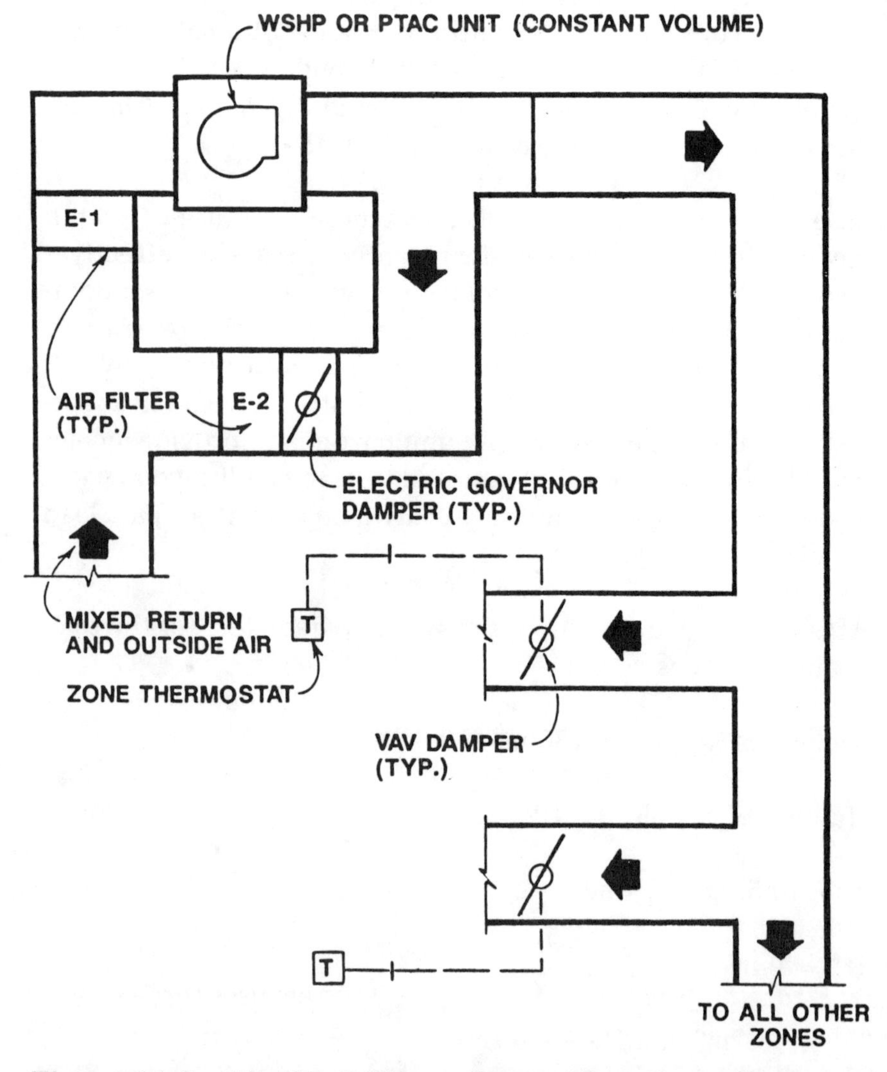

Figure 11-3. Modified Water Source Heat Pump/Packaged Terminal Air Conditioner System.

$F_r = 0.5$

$C_o = 5\ \mu g/m^3$

The combined filter effectiveness, E_f, above for parallel connection is calculated using equation (1) in Chapter 7 of Part 2, where X is the bypass percentage (11) and $E_1 = 0.5$ and $E_2 = 0.9$.

To satisfy $C_s < 50\ \mu g/m^3$ and $C_s < 150\ \mu g/m^3$ at long- and short-terms, respectively, and a minimum outside supply air of 15 cfm/person at part- as well as full-loads:

$V_o = 3185$ cfm with $C_s = 125\ \mu g/m^3$ and

$OSA_{calc.} = 15$ cfm/person at 50% part-load.

Notice that the required V_o of 3185 cfm for this system is less than that required (4358 cfm) for a conventional VAV system.

11.2.4 Case 4. Integrated Desiccant Cold Air Distribution System (Class VI in Table 11-1)

The integrated desiccant cold air distribution (IDCAD) system (11, 12) consists of an evaporative chilling (EC) module and a desiccant dehumidifier (DD) module integrated into a multi-duct, air-handling (MDAH) module, as shown in Figure 11-4. Each fully integrated dual-duct/VAV desiccant air-handling unit serves a combination of interior and exterior conditioned spaces by means of a unique three-deck air distribution system that takes full advantage of VAV energy-saving benefits for space cooling while maintaining satisfactory ventilation efficiency at variable part-load conditions by different means in the building's interior and exterior zones.

In Figure 11-4, notice that the IDCAD system uses three supply fans, one for the hot deck and two for each cold deck (operating at 40°F supply air temperature), thereby allowing for separation of exterior and interior zones while reducing return air duct-

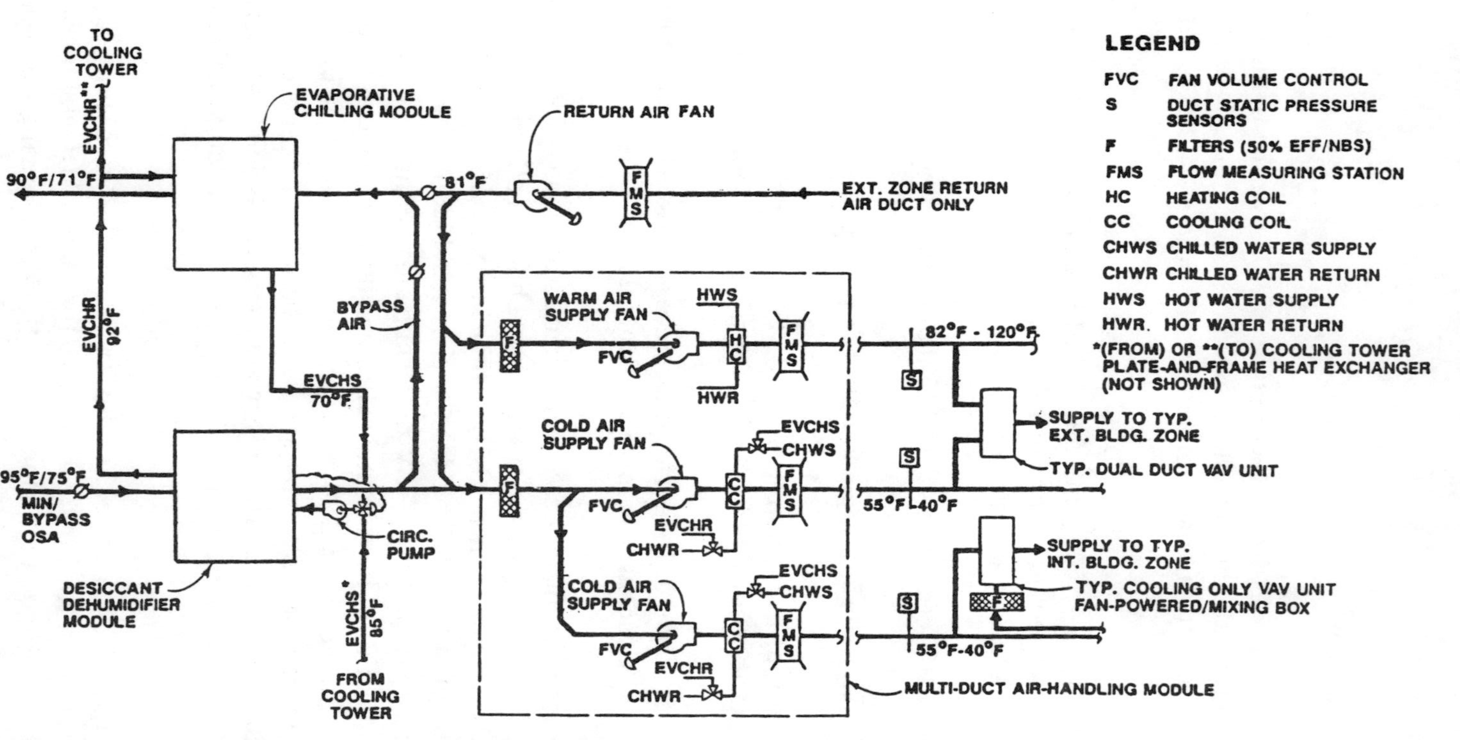

Figure 11-4. Integrated Desiccant Cold Air Distribution System.

work and separately monitoring for IAQ. Each supply fan's volume is controlled independently by the static pressure in its respective duct. Common return fan airflow is controlled relative to the sum of the hot and cold fan volumes using flow measuring stations to ensure positive tracking.

Each supply fan is sized for the anticipated maximum coincident hot or cold volume, not the sum of their instantaneous peaks. Each cold deck can be maintained at a constant temperature either by operating a cooling coil as a wet economizer with only minimum outdoor air (instead of a more costly free-cooling air economizer) when outside air is below the cold deck setpoint. Cold-air volumes and the refrigeration capacity are determined by sum of the simultaneous peaks of the cold-air requirements for all interior and exterior zones. Also notice that the return air fan is sized only for the returning exterior zone space air since the interior zones can use close-coupled, fan-powered VAV mixing boxes, representing major savings in operating cost, initial cost, and usable building space. Use of combination dual-duct VAV mixing boxes equipped with suitable terminal air cleaners permits higher recirculation rates and lower outdoor air ventilation rates without sacrificing IAQ.

At 100% full-load condition, the IDCAD system parameters for offices zone occupancy are:

Volume of space = 2039 m^3 (72,000 ft^3)

Interior = 6000 ft^2

Exterior = 3000 ft^2

Number of people = 63

$\dot{N} = 5\ \mu g/min \cdot m^3$

$E_V = 0.8$

$V_S = 210\ m^3/min$ (7430 cfm)

$E_f = 0.5$

$F_r = 1$

$C_o = 1\ \mu g/m^3$

To satisfy $C_s < 50\ \mu g/m^3$ and $C_s < 150\ \mu g/m^3$ at long- and short-terms, respectively, and a minimum outside air of 15 cfm/person at part- as well as full-loads:

$V_o = 975$ cfm with $C_s = 107\ \mu g/m^3$ and

$OSA_{calc.} = 15$ cfm/person at 100% full-load.

At 50% part-load condition (70% part-load for interior and 30% part-load for exterior zones), the system parameters for the interior offices zone are:

Volume of space = 1359 m^3 (48,000 ft^3)

Number of people = 29

$\dot{N} = 3.5\ \mu g/min \cdot m^3$

$E_v = 0.8$

$V_s = 105.2\ m^3/min$ (3715 cfm)

$E_f = 0.5$

$F_r = 1$

$C_o = 1\ \mu g/m^3$

The system parameters for the exterior offices zone are:

Volume of space = 680 m^3 (24,000 ft^3)

Number of people = 11

$\dot{N} = 2.5\ \mu g/min \cdot m^3$

$E_V = 0.8$

$V_s = 105.2\ m^3/min\ (3715\ cfm)$

$E_f = 0.5$

$F_r = 0.7$

$C_o = 1\ \mu g/m^3$

To satisfy $C_s < 50\ \mu g/m^3$ and $C_s < 150\ \mu g/m^3$ at long- and short-terms, respectively, and a minimum outside air of 15 cfm/person at part- as well as full-loads:

$V_o = 682$ cfm with $C_s = 102\ \mu g/m^3$ (interior zones),

$V_o = 292$ cfm with $C_s = 78\ \mu g/m^3$ (exterior zones), and

$OSA_{calc.} = 23$ cfm/person (interior) and 28 cfm/person (exterior).

Notice that C_o (in both part- and full-loads) is decreased to 1 since the DD module in Figure 11-4 utilizes a spray coil liquid desiccant dehumidifier to clean the outdoor air, thereby reducing C_o from initially 5 $\mu g/m^3$ to 1 $\mu g/m^3$. This system eliminates the need to provide outside air at full-load to satisfy part-load conditions of C_s seen in other systems.

Table 11-4 summarizes the results of the above four case studies listing the required outside supply air per person (cfm) and indoor contaminant level, C_s, as determined by the IAQ procedure. This contrasted with VR procedure for each of the four VAV systems. In Table 11-4, notice that if one simply relies on the VR procedure for example, shops occupancy and employs the conventional VAV system in Case 2, it is likely that the NPAAQS limits (150 $\mu g/m^3$) for the short-term would be exceeded, contrary to the intent of ASHRAE Standard 62-1989 (with respect to NPAAQS being exceeded indoors). The use of the IAQ pro-

ZONE	VENTILATION RATE PROCEDURE CFM/PERSON	INDOOR AIR QUALITY PROCEDURE CFM/PERSON (Cs, $\mu g/m^3$)							
		VAV SYSTEM WITH CONSTANT OSA		VAV SYSTEM W/PROP_ ORTIONAL OSA		MODIFIED WSHP/PTAC		IDCAD	
		100% LOAD	50% LOAD	100% LOAD	50% LOAD	100% LOAD	50% LOAD	100% LOAD	50% LOAD
TELECOMMUNICATIONS	20	15 (89)	21 (110)	20 (81)	15 (137)	15 (89)	15 (125)	15 (129)	19 (130)
CONFERENCE ROOMS	20	15 (93)	21 (109)	21 (84)	15 (137)	15 (93)	15 (129)	15 (133)	17 (140)
OFFICES	20	15 (65)	22 (135)	23 (63)	16 (147)	15 (65)	15 (121)	15 (107)	24 (94)
SHOPS	15	15 (78)	22 (114)	28 (74)	20 (140)	15 (78)	13 (124)	15 (118)	24 (110)

Table 11-4. Performance Summary of Four Variable- Air-Volume Systems at 100% Full- and 50% Part-Load Conditions for Various Zones Using Indoor Air Quality Procedure.

cedure as a means to validate the VR procedure results may be advisable to comply with the NPAAQS criteria. The parameters are calculated the same way as above at 100% full-and 50% part-load conditions.

If ventilation reduction in multiple zones supplied from a common source is required, it can be calculated using the graph in Figure 11-5 (13). For example, for the combination of communications and offices zones in a VAV system with constant temperature and proportional outdoor air (Class VII in Table 11-1), the ventilation reduction is calculated as follows:

$$X = \frac{4378 + 975}{12{,}740 + 13{,}000} = 0.21$$

$$Z = \frac{4358}{12{,}740} = 0.34$$

where X is the sum of all zone outdoor air flows divided by the total supply, and Z is the outdoor air fraction required in supply to a critical space. From the graph in Figure 11-5, the corresponding ventilation fraction, Y, is 0.24, which implies that the total outside air is 25,740 × 0.24 = 6180 cfm.

11.3 Lead/Lag Procedure

To provide adequate dilution of air to maintain contaminant concentrations within acceptable ranges of a ventilation system with variable occupancy, the outdoor air quantity must be adjusted either by dampers or stopping and restarting ventilation (13). This adjustment may lag or should lead occupancy, depending on the contamination source and the change in occupancy. When the contamination concentration is due to occupancy only, it is dissipated during unoccupied periods and supply of outdoor

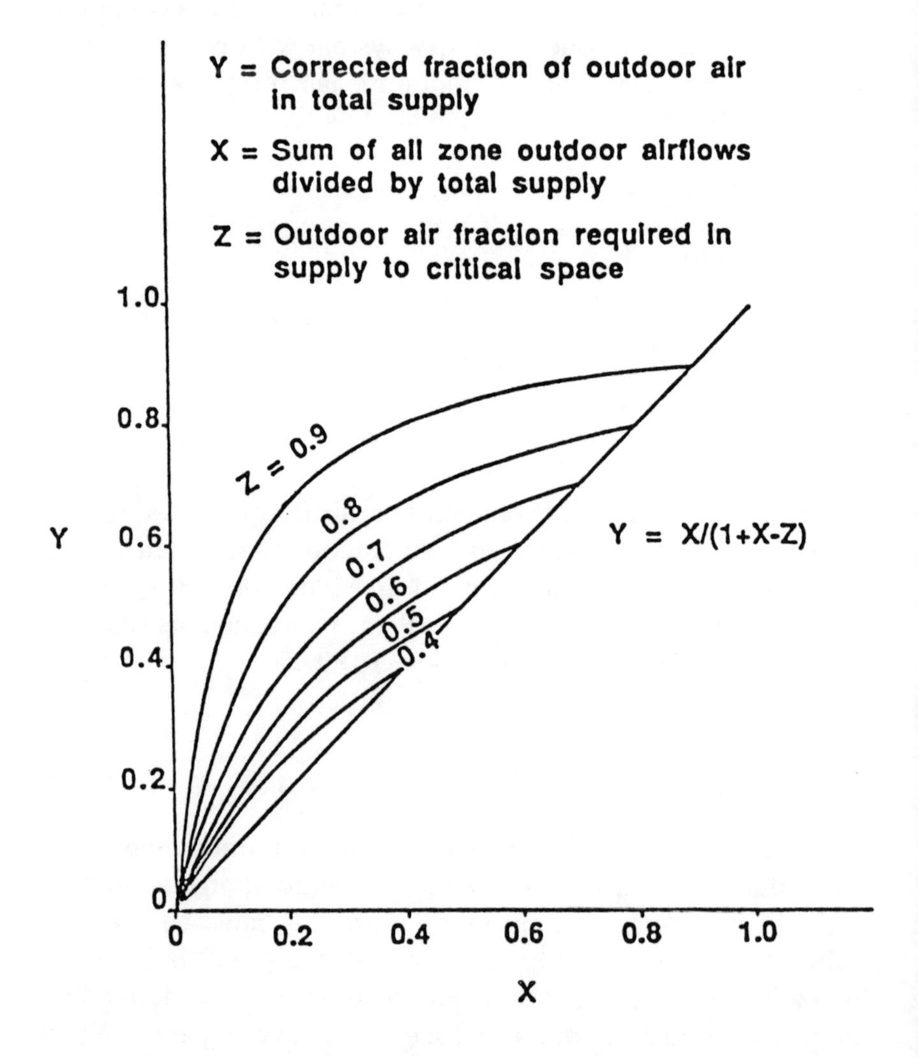

REFERENCE : ASHRAE Standard 62-1989

Figure 11-5. Ventilation Reduction Permitted in Multiple Zones Supplied from a Common Source.

air may lag occupancy. If the contaminant level in a zone is independent of occupancy, the outdoor air supplied must lead occupancy to have acceptable concentrations at the start of occupancy. Figures 11-6 and 11-7 illustrate the lag and lead times, respectively, necessary to obtain acceptable concentrations for variable occupancy.

The following two examples illustrate lag and lead times for the telecommunications zone and conference rooms zone. For the conference rooms zone, the contaminant level is assumed to be due to occupancy, and the contaminant level for the telecommunications zone is assumed to be independent of occupancy due to equipment-related contamination.

For the 32,000 ft^3 conference rooms zone and calculated outside supply air of 15 cfm/person in Case 1, the air capacity in the space is 160 ft^3/person. Then, from the graph in Figure 11-6 (13), the corresponding maximum permissible ventilation lag time is 0.15 hour or 9 minutes.

By the use of the same procedure as above, for the 40,000 ft^3 telecommunications zone and calculated outside supply air of 15 cfm/person (also in Case 1), the air capacity in the space is 133 ft^3/person. Then, from the graph in Figure 11-7 (13), the corresponding lead time is 0.6 hour or 36 minutes. This implies that outdoor air supplied leads occupancy to have acceptable concentrations at the start of occupancy.

11.4 Summary

Since a large amount of our time is spent indoors, IAQ is a very important factor to our health and comfort. IAQ directly affects the energy consumption of a building due to HVAC system ventilation air requirements and air distribution characteristics. For example, conditioning and distribution costs associated with moving air in buildings account for 50% to 60% of the nation's total building energy consumption estimated at more than $80 billion

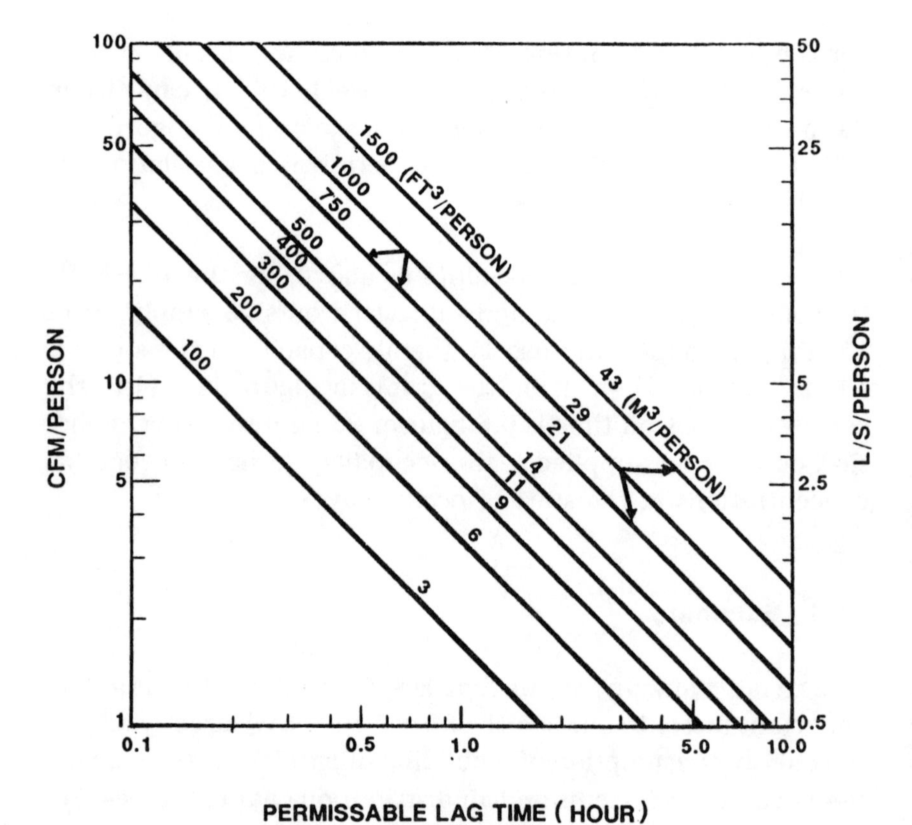

REFERENCE : ASHRAE Standard 62-1989

Figure 11-6. Maximum Permissible Ventilation Lag Time.

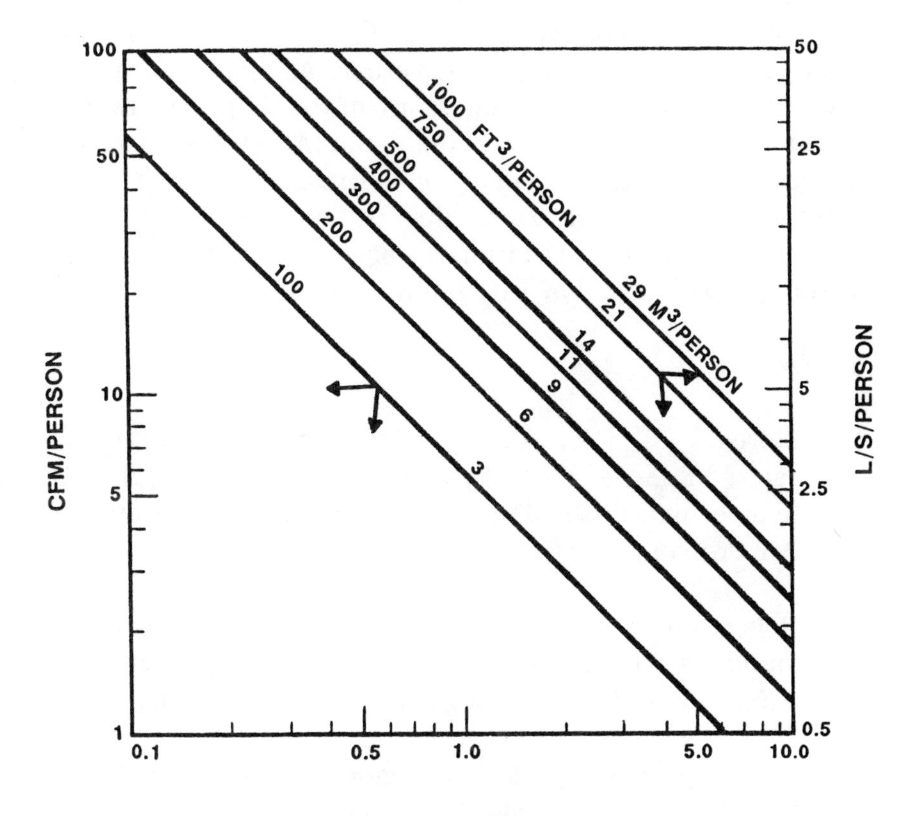

REFERENCE : ASHRAE Standard 62-1989

Figure 11-7. Minimum Ventilation Time Required Before Occupancy of Space.

per year. If building ventilation is improved, not only could occupants have better IAQ, but building owners could also save up to $2 billion a year (14).

The above examples examined four different VAV system options for a representative office building located in metropolitan Los Angeles in SCAQMD. Applying the latest SCAQMD data, it would appear that the following outdoor air pollutants, particulates (PM 10), CO, and ozone concentrations have consistently exceeded the limits with respect to existing state and federal standards set by the SCAQMD within the last six months. Where such information exists (e.g., in the case of shops occupancy at 50% part-load [refer to Table 11-4 for Case 2]), optional use of the VR procedure would result in excess C_s of 150 $\mu g/m^3$ for short-term occupancy, as pointed out earlier. As more information becomes available, indiscriminate use of the VR procedure may have to be reexamined. Accordingly, the IAQ procedure can be utilized to verify the adequacy of the results obtained by the VR procedure as well as to more directly compute indoor air contaminant concentrations within the occupied building space applying the techniques of IAQ zoning.

References

1. Morey, P.R. "Microorganisms in Buildings and HVAC Systems: A Summary of 21 Environmental Studies." ASHRAE Symposium, Proceedings: Engineering Solutions to Indoor Air Problems - IAQ '88, Atlanta, GA, 1988.

2. Robertson, G. "Ventilation, Heating and Energy Conservation - A Workable Compromise." Proceedings of the 11th Annual World Energy Engineering Congress, 1988.

3. South Coast Air Quality Management District (SCAQMD) "Air Quality Standards Compliance Report." Vol. 1, No. 1-6, 1988.

4. Meckler, M. "Ventilation/Air Distribution Cure Sick Buildings." *Specifying Engineer*, 1985.

5. Swedish Council for Building Research. "Analysis of Low Particulate Size Concentration Levels in Office Environments." Proceedings of the 3rd International Conference on Indoor Air Quality and Climate, Vol. 2, Stockholm, 1984.

6. Meckler, M. "Computer Design, Monitoring, Control and Modeling of HVAC Systems." *Sol*AIR*, Vol. 30, No. 5, 1986.

7. Meckler, M. "Off-Peak Desiccant Cooling Combine to Maximize Gas Utilization." *ASHRAE Transactions*, Vol. 94, Pt. 1, 1988.

8. Meckler, M. "Multisource Hydronic Heat Pump System Cuts Energy Costs." *Consulting-Specifying Engineer*, 1987.

9. Meckler, M. "Indoor Air Quality Vs. Energy Efficiency: Impact of New Ventilation Standards." *Consulting-Specifying Engineer*, 1988.

10. Meckler, M. and Janssen, J. "Use of Air Cleaners to Reduce Ventilation." ASHRAE Symposium, Proceedings: Engineering Solutions to Indoor Air Problems - IAQ '88, pp. 130-147, Atlanta, GA, 1988.

11. Meckler, M. "Integrated Multi-Duct Dual-Stage Dual-Cooling Media Air Conditioning." U.S. Patent 4,300,623, 1981.

12. Meckler, M. "Integrated Desiccant Cooling/TES/Cogeneration Systems." Proceedings of the Utilization Engineering/Marketing Symposium, American Gas Association, Baltimore, MD, 1988.

13. ASHRAE Standard 62-1989, "Ventilation for Acceptable Indoor Air Quality." Atlanta, GA, 1989.

14. Solar Energy Research Institute. "Indoor Air Quality." Golden, CO, 1988.

12. SOLID DESICCANTS AND INDOOR AIR QUALITY CONTROL

Demetrios J. Moschandreas, Ph.D.
Sr. Science Advisor

and

Seresh Relwani
Research Engineer
IIT Research Institute
Chicago, Illinois

Page

12.1 Introduction

When the effect of desiccant dehumidification is considered two issues must be addressed: (a) can desiccants be used to remove contaminants from indoor environments? and (b) to what extent does humidity interfere with effectiveness of desiccants to remove indoor air pollutants? Several attempts have been made to study the capabilities of molecular sieve, silica gel, and other desiccants to reduce the indoor concentrations of carbon monoxide (CO), nitrogen dioxide (NO_2), sulfur dioxide (SO_2), and total hydrocarbons (THC).

Adsorption is employed commercially for the separation and purification of a wide variety of gas and liquid streams. The commercial manufacturing of molecular sieves has provided the primary impetus for the enhanced use of adsorption processing. The physical and chemical principles used for control of indoor air pollution are the same as those used for control of industrial effluents. The most marked difference relates to the contaminant to be controlled: indoor air contaminant concentrations are lower than industrial concentrations. This chapter presents a brief review of experiments specifically designed to address indoor air quality (IAQ) using solid desiccants.

It is assumed that the reader is familiar with concepts associated with adsorption including the mass transfer zone (MTZ), breakthrough time, log time, and the sharpness. In the following paragraphs, a brief qualitative summary of adsorption based on research (1) is provided.

Although the theoretical difference between the physical and chemical adsorption is clear in practice, the distinction is not as simple as it may seem. The following parameters can be used to evaluate an adsorbate-adsorbent system to establish the type of adsorption:

- The heat of physical adsorption is in the same order of magnitude as the heat of liquefication, while the heat of chemisorption is of the corresponding chemical reaction.

- Physical adsorption will occur under suitable temperature and pressure conditions in any gas-solid system, while chemisorption takes place only if the gas is capable of forming a chemical bond with the surface.

- A physically adsorbed molecule can be removed unchanged at a reduced pressure at the same temperature where the adsorption took place. The removal of the chemisorbed-layer is far more difficult.

- Physical adsorption can involve the formation of multi-molecular layers, while chemisorption is always completed by the formation of a monolayer. In some cases physical adsorption may take place on top of a chemisorbed-monolayer.

- Physical adsorption is instantaneous (diffusion into porous adsorbents which is time-consuming) while chemisorption may be instantaneous, but generally requires an activation energy.

The boundary layer is the most important factor in the phase interaction. Therefore, to achieve a high rate of adsorption, it is expedient to create the maximum obtainable surface area within the solid phase. High surface area can be produced by creating a large number of microcapillaries in the solid. All commercial adsorbents, such as activated carbon, silica gel, alumina, etc., are prepared in this manner.

Figure 12-1 shows five basic types of adsorption isotherms (2). The Type I isotherm represents systems in which adsorption does not proceed beyond the formation of a monomolecular layer. The Type II isotherm indicates an indefinite multilayer formation after the completion of monolayer. For example, the adsorption of water vapor on carbon black at 30 °C results in such a curve. The Type III isotherm is obtained when the amount of gas adsorbed increases without a limit as its relative saturation approaches unity. The convex structure is caused by the heat of adsorption of the first layer becoming less than the heat of condensation due to molecular interaction in the monolayer. This type of isotherm is obtained when adsorbing bromine on silica gel at 20 °C. The Type IV isotherm is a variation of Type II, but with a finite multilayer formation corresponding to complete filling of the capillaries. This type of isotherm is obtained by the adsorption of water vapor on active carbon at 30 °C. The Type V isotherm is similar to Type III. For example, it is obtained when adsorbing water vapor on activated carbon at 100 °C.

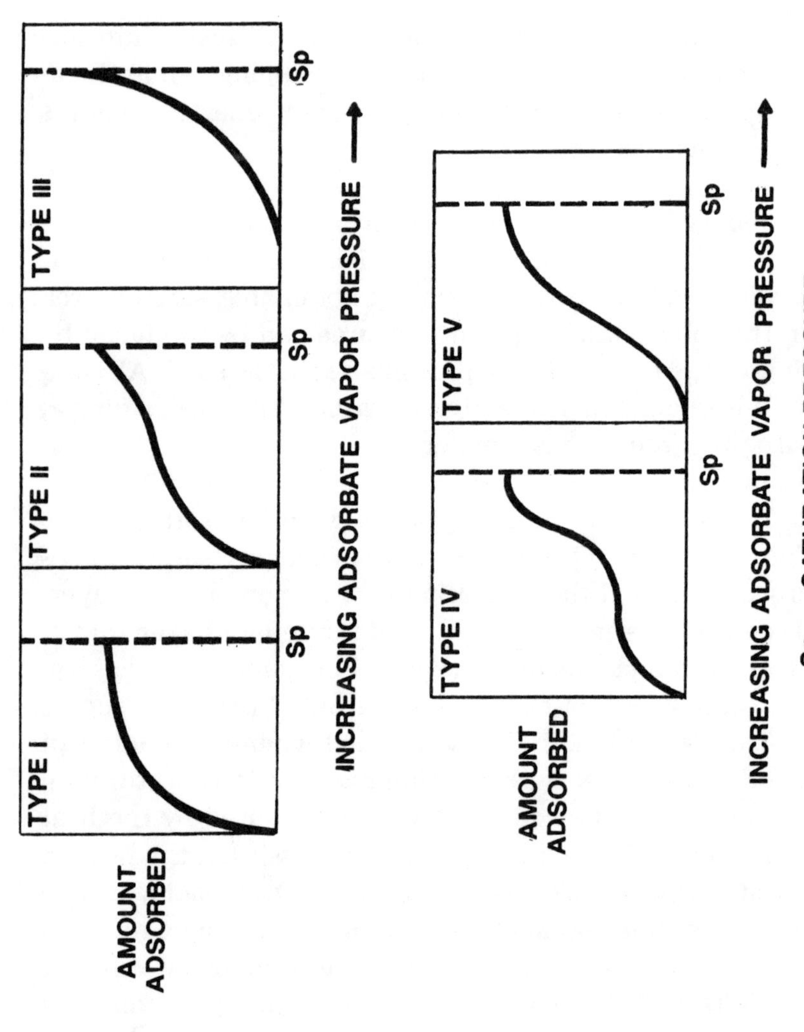

Figure 12-1. Types of Adsorption Isotherms.

Although a great number of equations have been developed to date based on theoretical considerations, none of them can be generalized to describe all systems (2). Isotherms are indicative of the efficiency of an adsorbent for a particular adsorbate removal, yet they do not supply data to enable the calculation of contact time nor the amount of adsorbent required to reduce the contaminant concentration below the required limits.

The following factors play the most important role in dynamic adsorption and the length and shape of the MTZ:

- The type of adsorbent,
- The particle size of an adsorbent (may depend on a maximum allowable pressure drop),
- The depth of the adsorbent bed,
- The gas velocity,
- The temperature of the gas stream and the adsorbent,
- The concentration of the contaminants to be removed,
- The concentration of the contaminants not to be removed (including moisture),
- The pressure of the system,
- The removal efficiency required, and
- Possible decomposition of contaminants on the adsorbent.

Most adsorbents are capable of adsorbing both the organic and inorganic gases. However, their preferential adsorption characteristics and other physical properties make each one more or less specific for a particular application. For example, activated alumina, silica gel, and molecular sieves will adsorb water preferentially from a gas-phase mixture of water vapor and an organic contaminant. This is a considerable drawback in the application of these adsorbents for removal of organic contaminants. Activated carbon preferentially adsorbs nonpolor organic compounds. Recently, through evaluation of the type of surface oxides present on activated carbon surfaces, it was found that the preferential adsorption properties of carbon can be partially regulated by the type of surface oxide induced on the carbon.

In some cases, one of the adsorbents has sufficiently retained adsorption capacity for a particular contaminant. In these applications, a large surface area adsorbent is impregnated with an inorganic or, in rare cases, with a high-molecular-weight organic compound which can chemically react with the particular contaminant. Bromine-impregnated carbons are used for removal of ethylene or propylene (3). The action of these impregnants is either by catalytic conversion or reaction to a nonobjectionable compound, or to a more easily adsorbed compound. An impregnated adsorbent is available for most compounds which, under particular conditions, are not easily adsorbed by nonimpregnated commercial adsorbents.

Adsorption takes place at the interphase boundary, therefore, the surface area of the adsorbent is an important factor in the adsorption process. Generally, the higher the surface area of an adsorbent, the higher its adsorption capacity is for all compounds. However, the surface area has to be available in a particular pore size within the adsorbent. At low partial pressures (concentration), the surface area in the smallest pores in which the adsorbate can enter is the most efficient. At higher pressures, larger pores become more important, while at very high concentrations, capillary condensation takes place within the pores and the total micropore volume is the limiting factor. The most valuable information concerning the adsorption capacity of a certain adsorbent is its surface area and pore volume distribution curve in different diameter pores. The relationship between the adsorption capacity and surface area in optimum pore sizes depends on concentration. Therefore, it is very important that any evaluation of adsorption capacity is performed under actual concentration conditions.

The action of molecular sieves is slightly different from those of other adsorbents in that selectivity is determined by the pore size limitations of the molecular sieve. In selecting molecular sieves, it is important that the contaminant to be removed be smaller than the available pore size, while the carrier gas of the component (not to be removed) is larger, thus, not adsorbed. Because the optimum pore size varies with concentration, molecular sieves are limited in their use by the applicable concentration ranges.

12.2 Desiccants and Contaminant Adsorption

A summary of known measurements for SO_2 adsorption on zeolites is presented in Table 12-1 (4). Presently, kinetic data necessary to develop or design a large-scale adsorber does not exist. Aside from the work of Vinnikov (5), the existence of kinetic data for the simultaneous removal of SO_2 and water vapor was not found.

The results for the binary adsorption tests on 13X molecular sieve are summarized in Table 12-2 (4). The measured breakthrough curves for different SO_2 water-vapor-mixture fractions are shown in Figure 12-2 (4). Figure 12-2 shows the results for a high SO_2 water-vapor-mixture fraction (5.28 ppmv H_2O/ppmv SO_2). The interesting phenomenon to notice in Figure 12-2 is that the effluent SO_2 peak actually exceeds the inlet concentration. In fact, the effluent peak represents a desorption of 18.5% of the initial SO_2 adsorbed. The reason is that water vapor is more strongly attracted and held by adsorbent than SO_2. Consequently, water vapor is adsorbed in the first few adsorbent layers of the bed. As the water vapor MTZ moves through the bed, the adsorbed SO_2 is displaced and consequently raises the free stream SO_2 concentration. The SO_2 adsorbed was 6.38 g/100g adsorbent compared to 16.1 g/100g for the water vapor free case.

Macriss, Rush, and Well (6) used an improved adsorbent system which employs thin sheets of layers of a fibrous material impregnated with 10% to 90% by weight of a solid adsorbent comprising of finely powdered, solid, crystalline alkali metal or alkaline earth metal alumino silicates that have the water of hydration removed. The adsorbent materials particularly useful are crystalline natural or synthetic zeolites.

Adsorbent	SO_2 Concentration	Composition of Carrier				SO_2 Adsorbed	Temp.	Equilibrium Time	Invest.
	Vol. pct.	N_2	CO_2	O_2	H_2O	g/100g	°C	min.	
(Molecular Sieves)									Martin and
Linde 13X	9	81	10	-	-	27.6	26	70	Brantley
	3	87	10	-	-	23.9	26	-	
	1	97	10	-	-	22.2	26	-	
Linde 5A	9	81	10	-	-	18.9	26	41-45	
Linde AW500	9	81	10	-	-	18.4	26	50	
	3	87	10	-	-	16.5	26	-	
	1	89	10	-	-	14.2	26	-	
Linde 13X	9	81	10	-	-	29.5	26	35-37	
	3	87	10	-	-	27.5	26	-	
	1	89	10	-	-	25.5	26	-	
Davison 13X	0.35	82.7	14.9	-	-	6	125	-	Friedman
Norton--N(cation)	0.35	82.7	14.9	-	-	1-2	125	-	
Norton--Cu(cation)	0.35	82.7	14.9	-	-	1-2	125	-	
Norton--Ni(cation)	0.35	82.7	14.9	-	-	1-2	125	-	
Norton--N(cation)	0.35	82.7	14.9	-	-	3.9	23	-	
Natural Mordenite	100	-	-	-	-	10.223	28	150	Tamboli
Linde AW300	100	-	-	-	-	9.721	28	100	
Linde 4A	0.3	74.7	13.0	6.0	6.0	2.3	130	-	Bienstock
Linde 5A	0.3	74.7	13.0	6.0	6.0	1.8	130	-	and Field
Linde 13X	0.3	74.7	13.0	6.0	6.0	7.0	130	-	
Zeolite (NaX)	0.37	78.0	21.0	0.45	-	2.2	-	-	Vinnikov
Zeolite N-Mordenite	0.37	78.0	21.0	0.45	-	1.8	-	-	
(Silica gel)									
Gel (ShSM)	0.37	-	-	-	0.45	1.0	-	-	
Gel (ShSM)	0.59	-	-	-	0.01	2.2	-	-	
Gel (ShSM)	0.30	-	-	-	2.5	2.0	-	-	
						cc/g			
Silica Gel (Labora-	110	-	-	-	-	110	25	-	McGavack
tory sample)	100	-	-	-	-	25	100	-	and Patrick

Table 12-1. Sulfur Dioxide Adsorption on Molecular Sieves and Silica Gel.

Adsorbent	Gas Temperature °C(°F)	Gas Velocity m/s (ft/s)	SO_2 Concentration ppmv	H_2O Concentration ppmv	Effluent Time at Half-Maximum t_{SO_2}	t_{H_20}	SO_2 Adsorption g/100g	SO_2 Desorption Percent
13X	26 (80)	0.34 (1.1)	2480	13,100	59.5	113	6.38	19
	23 (75)	0.33 (1.1)	2800	3,960	87.5	231	10.2	41
	60 (140)	0.89 (2.9)	2800	14,900	10.5	38.5	4.54	8.6
	77 (170)	0.96 (3.2)	2800	11,400	14	42	6.3	0.2
AW500	23 (73)	0.34 (1.1)	2930	4,180	36	240	5.3	17
	29 (85)	0.34 (1.1)	2660	12,700	24.5	98	2.6	51
	77 (170)	0.96 (3.2)	3000	11,300	7	32	1.4	29
	31 (88)	0.86 (2.8)	2680	9,990	7	45.5	1.7	15
Silica Gel	25 (78)	0.018 (0.06)	2630	10,600	1.8	35	0.02	2.1
	34 (93)	0.86 (2.8)	3150	13,800	5.3	21	0.66	4.8
	25 (78)	0.85 (2.8)	2750	5,300	7	47	1.8	47

Table 12-2. Summary of Binary Adsorption Tests.

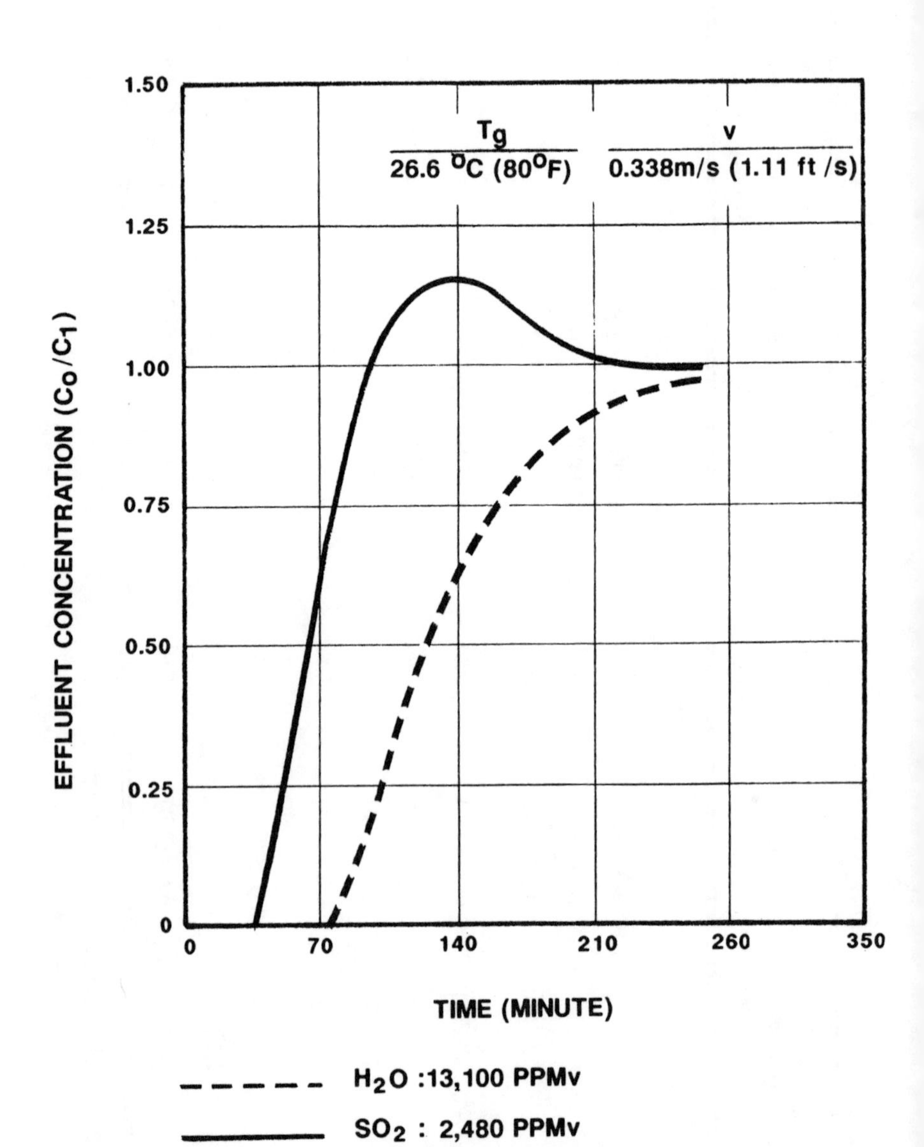

Figure 12-2. Binary Adsorption Performance of SO_2 and Water Vapor on 13X Molecular Sieve-High Water Vapor Content.

It is the objective of this invention to provide a process and apparatus for removal of NO_2, SO_2, and CO_2 from environmental air by adsorption on molecular sieves in a continuous and adsorbent regenerable system. In their experiments (6) they found that when ambient air at 80 °F, absolute humidity of 0.017 lb/lb of dry air and 0.5 ppm NO_2, is introduced into the system, the air leaves the system with less than 0.05 ppm NO_2.

Alumina-supported cobalt, chromium, and manganese sorbent were shown by Gidaspow (7) to adsorb NO_x in the temperature range of 25 °C to 45 °C in the presence of oxygen and water vapor.

Co-precipitated iron oxide-cobalt oxide sorbent containing sodium was developed (7) that has a capacity 20 times that of the best supported transition metal oxides. Water vapor normally present in the flue gas did not appear to decrease the capacity of the sorbent. However, further investigation using simulated flue gases to make certain that no chemical interference will occur, is needed.

Reaction rate data are analyzed using previously developed models. Data can be correlated successfully by taking into account diffusion of nitrate oxide into the particle simultaneously with its reaction with an active sorbent.

The adsorption of low concentrations of n-decane, methane, benzene, toluene, xylene, pyridine, aniline, and nitrobenzene on type 13X Linde molecular sieve has been studied at influent concentrations of 5 ppm to over 100 ppm and linear velocities of up to 2000 cm/minute for a representative aliphatic compound (n-decane) was 10 weight-percent (wt-%) of the sieve and for a mononuclear aromatic (benzene) was 0.35 wt-% at 25 °C. Decreasing the flow velocity of the aromatic to 500 cm/minute increased the breakthrough capacity to 2.1%.

When co-adsorption of n-decane and H_2O takes place on 13X sieve, in Figure 12-3 (8), the wt-% of n-decane sorbed at breakthrough drops sharply below what is sorbed from dry air (8). For example, at an n-decane concentration of 100 ppm in air with 48% relative humidity (RH), the organic vapor loading drops to 1 wt-% from 10.1 wt-% with dry air; at the same time 13.5 wt-% of water is adsorbed. This is to be expected since the concentration of organic vapor is small relative to that of water. A sixfold increase in the n-decane concentration (with an RH of 48%) increases the breakthrough capacity to 4.4 wt-%; the weight of water co-adsorbed is 10 wt-%.

The above relationship does not hold true for co-adsorption of CO_2 and H_2O on molecular sieve. For example, the type 5A molecular sieve (refer to Table 12-1) at 20 °C adsorbed 4.4 wt-% CO_2 at breakthrough from dry air containing 1.5% CO_2. At 50% RH (H_2O = 8.76 mm of Hg), 3.3 wt-% CO_2 and 4.38 wt-% H_2O were adsorbed and at 100% RH (H_2O = 17.53 mm of Hg), 2.2 wt-% CO_2 and 8.76 wt-% H_2O were adsorbed.

The type 13X molecular sieve will adsorb at room temperature (25 °C to 35 °C) about 2% of its weight of benzene before breakthrough at 5 ppm and a gas velocity of 500 cm/minute. Increasing the flow velocity to 2500 cm/minute decreases the breakthrough capacity to 0.35 wt-%. The equilibrium capacity at this flow velocity is 4.5 wt-%. The adsorption is isothermal at these low organic vapor concentrations, if water is not present.

The breakthrough capacity for benzene at 100 ppm, 500 cm/minute and 25 °C is 7.5%. Increasing the linear velocity to 2500 cm/minute decreases the breakthrough capacity to 1.9%. Decreasing the influent temperature from 30 °C to 20 °C does not increase the breakthrough capacity of the sieve by more than 0.2%, but it will increase the equilibrium capacity to about 1.5%

The molecular sieve capacity for toluene at breakthrough is nearly the same as for benzene. The breakthrough capacity, at 25 °C, 2500 cm/minute linear velocity and an influent concentration of 3.5 ppm, is 0.4%. The equilibrium capacity at 3 ppm is 4.25 wt-%, while for benzene the equilibrium capacity at 3 ppm is 2.8 wt-%.

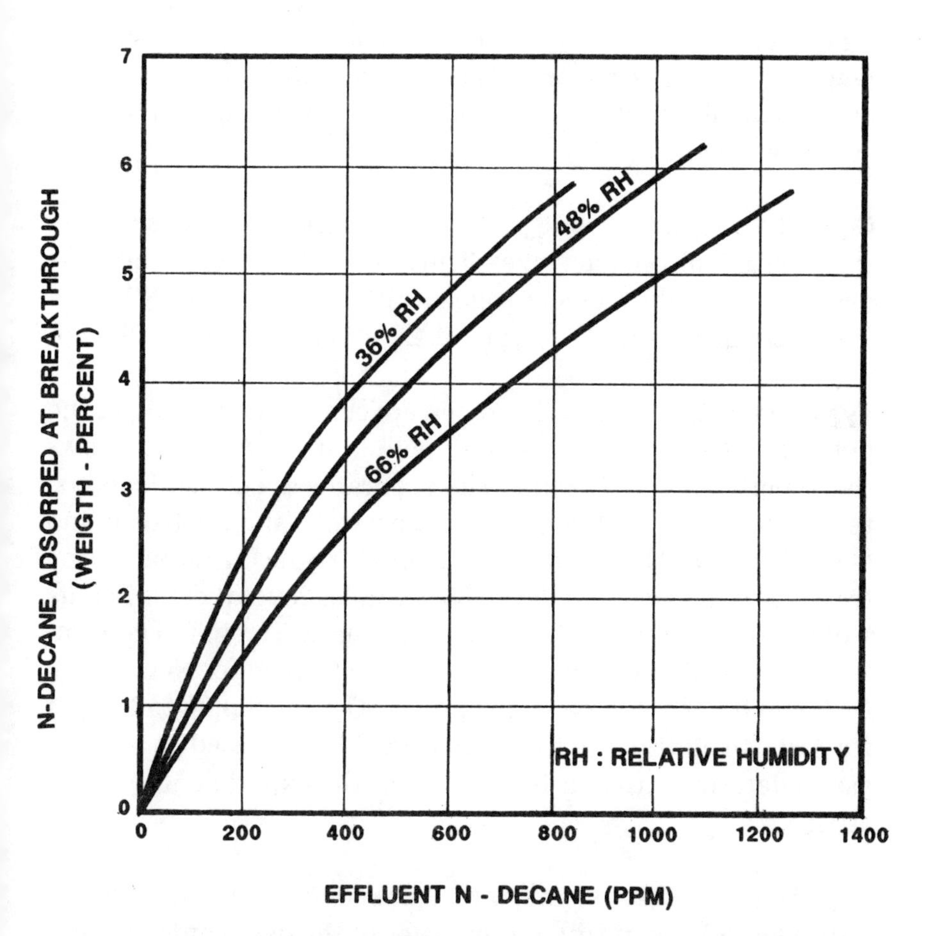

Figure 12-3. Effect of Relative Humidity on the Relationship Between Breakthrough Capacity of 13X Molecular Sieve and Breakthrough Concentration of N-Decane at 20°C and Flow Rate of 2500 cc/minute.

The results suggest that the high capacity iron oxide-cobalt oxide sorbent system in which the sorbent is incorporated in a thin sheet or fibrous material (7) is promising.

12.3 Indoor Air Quality and Desiccant Adsorption Laboratory Experiments

In the experiments (9) conducted using desiccants to reduce indoor concentrations of NO_2, CO, SO_2, and THC, three parameters were evaluated: (a) effect of air constituent concentration level, (b) desiccant materials, and (c) thermal regeneration capabilities of desiccants. It is concluded that desiccants can be used as a component of IAQ control systems.

Control experiments were performed using two solid desiccants: silica gel and molecular sieves. The removal capabilities of these desiccants to reduce indoor concentrations of NO_2, CO, SO_2, and THC were measured. Two types of experiments were performed.

Type I, bench-scale experiments, were conducted to explore the potential of desiccants to adsorb pollutants. Figure 12-4 shows the experimental apparatus. The apparatus is a closed-loop system to maximize the contact time between the airstream and desiccant bed. The decrease in pollutant concentration in the test loop was measured every 30 seconds. Several experiments were performed for each desiccant material. The initial concentrations of pollutants were 0.30 ppm to 0.50 ppm for NO_2, 2 ppm to 3 ppm for CO, 20 ppb to 50 ppb for SO_2, and 8 ppm to 12 ppm for THC. In addition to the pollutants, SF_6 was used as a tracer gas to determine the air infiltration of the experimental apparatus.

Type II, pilot-scale experiments, were conducted to define specific contaminant sorption properties of the desiccants and also to establish the fullness of contaminant release upon their thermal regeneration. A commercially available desiccant dehumidifier unit was used. The experimental apparatus shown in Figure 12-5 was built to challenge the desiccant material with CO, NO_2,

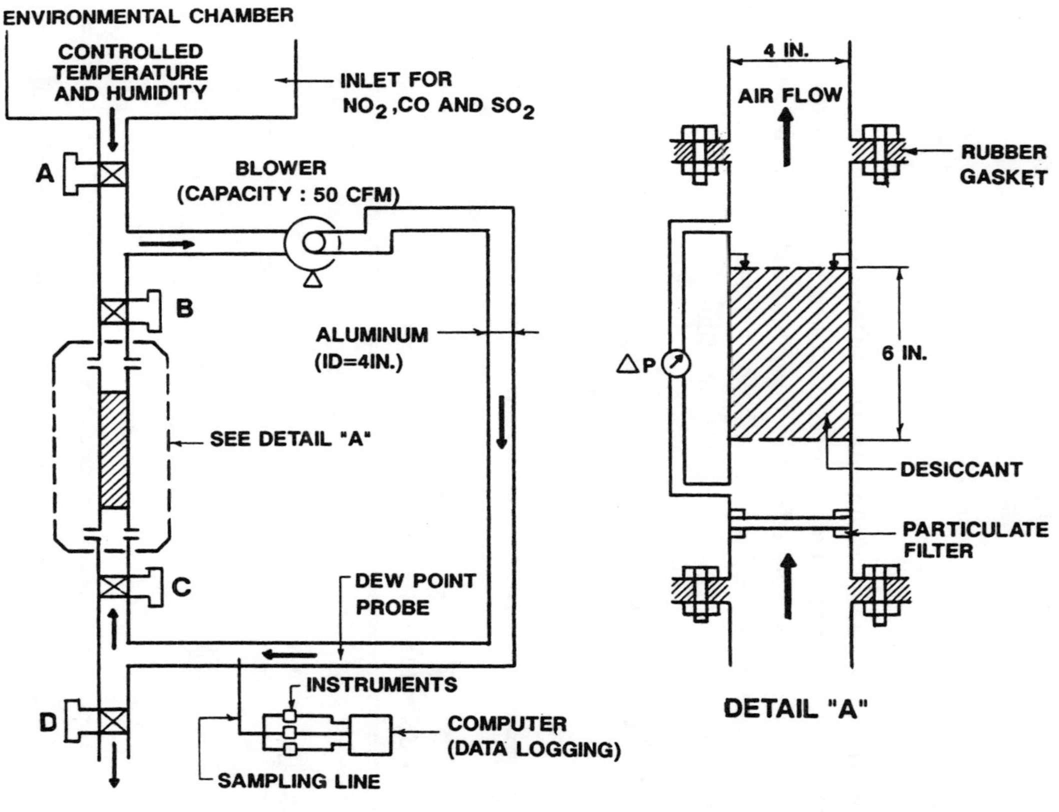

Figure 12-4. Experimental Setup for Type I Experiments.

and SO_2. Constituents were injected on the process air intake side to achieve the desired inlet conditions. Contaminant concentrations were monitored at four sampling locations: two upstream of the bed and two downstream of the bed. The sampling sequence and data logging were computer-automated. The dynamic leakage rate of the system was estimated by replacing the desiccant medium with nonreactive glass beads. A total of 60 experiments were conducted for several environmental conditions.

During the Type I experiments, the difference between the SF_6 and constituent removal rates is assumed to be the adsorption rate of the constituent for the desiccant material. Table 12-3 shows the percent reduction for both desiccant materials of each constituent for a 12-minute contact time.

Table 12-3. Percent Removal Rate of Each Constituent Estimated from Type I Experiments.

	Percent Reduction[A]	
Constituent	Molecular Sieve[B]	Silica Gel[C]
NO_2	36 ± 2	32 ± 2
CO	39 ± 1	43 ± 6
SO_2	46 ± 14	42 ± 11
THC	38 ± 12	56 ± 16

A: For a contact time of 12 minutes in closed-loop. Values are corrected for leakage rate.

B: Two experiments were performed for each constituent.

C: Two experiments were performed for each constituent except for THC, for which three were performed.

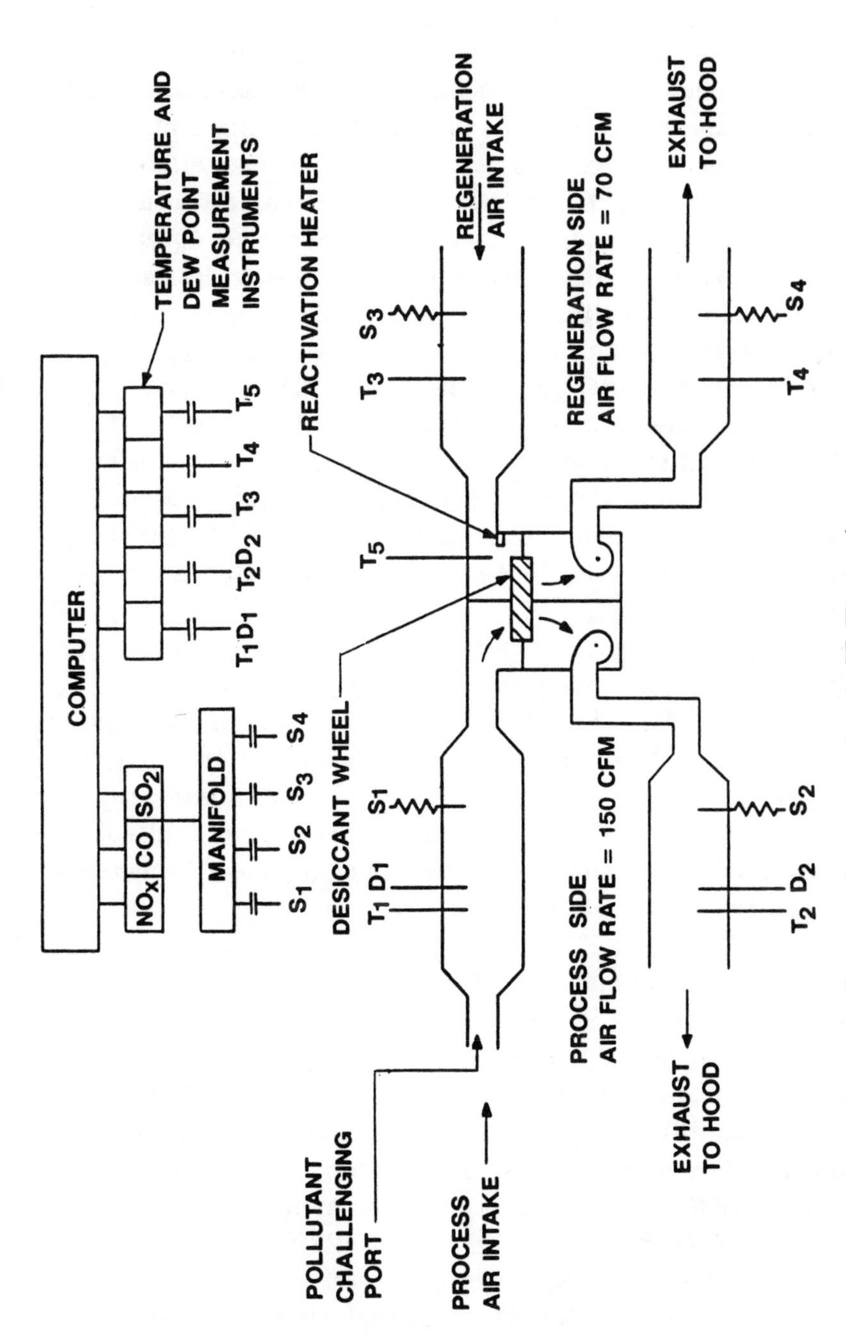

Figure 12-5. Experimental Setup for Type II Experiments.

During the Type II experiments, the removal efficiency of the desiccant bed and its thermal regeneration rate were determined from the ratio of the pollutant concentration in the air downstream of the desiccant bed to the corresponding value from the air upstream of the bed. The leakage rate of the unit and the cross leakage rate between the process and regeneration sides were estimated and corrected before determining both the removal and regeneration efficiencies of the desiccant materials. Table 12-4 shows the removal efficiency of both desiccants for each constituent. Different regeneration air temperatures were studied. For silica gel, low temperatures were in the range of 91 °C to 100 °C and high temperatures 134 °C to 232 °C. For molecular sieves, low temperatures were 106 °C to 109 °C and high temperatures 220 °C to 312 °C. The thermal regeneration capabilities of each desiccant material varied substantially with respect to contaminant, contaminant inlet concentration to the regeneration side, and the regeneration air temperature used.

The adsorption of gases onto solid surfaces is a well-known surface phenomenon. The adsorption may be either a chemisorption or physical adsorption. A combination of chemisorption on the surface, followed by a physical adsorption on top, is also possible. Although the mechanisms by which desiccants remove indoor air pollutants are not yet clear, the results presented here indicate that desiccants do indeed reduce the indoor concentrations of the pollutants tested.

The removal efficiency for NO_2, CO, and SO_2 discussed in this chapter are limited by the design parameters of the desiccant bed(s) and do not represent the optimum or best performance of the two desiccants. For both silica gel and molecular sieves, the removal efficiency for CO seemed to decrease as the inlet concentration increased, while the reverse trend was observed for NO_2 and SO_2. Molecular sieves demonstrated higher adsorption capabilities than silica gel for NO_2 (22% to 43% vs. 15% to 33%), for CO (10% to 61% vs. 10% to 36%), and for SO_2 (73% to 95% vs. 10% to 16%). The fact that molecular sieves have smaller pores than silica gel can partially explain this phenomenon. Adsorption takes place at the interphase boundary; there-

		Adsorption Capabilities					
		Silica Gel			Molecular Sieve 5A		
Air Constituent	Concentration Level[A]	No. of Exp.	Removal Efficiency[B] %	Absolute Reduction[B] $\mu g/m^3$	No. of Exp.	Removal Efficiency[B] %	Absolute Reduction[B] $\mu g/m^3$
CO	Low	4	21-36	115-279	3	45-61	187-427
CO	Medium	4	17-32	268-437	4	19-39	301-414
CO	High	4	10-18	411-1871	4	10-22	499-957
NO_2	Low	3	16-23	13-22	3	22-36	16-21
NO_2	Medium	4	15-25	35-160	2	20-38	173-404
NO_2	High	4	19-33	203-406	2	43	485-541
SO_2	Low	1	10	11	5	73-82	29-69
SO_2	Medium	5	9-17	16-67	4	82-94	93-336
SO_2	High	3	16-18	90-132	1	95	468

A: For CO: Low = 550-1000 $\mu g/m^3$, Medium = 1000-2000 $\mu g/m^3$, High = 1300-2000 $\mu g/m^3$

For NO_2: Low = 50-100 $\mu g/m^3$, Medium = 100-1000 $\mu g/m^3$, High = 1000-2000 $\mu g/m^3$

For SO_2: Low = 50-110 $\mu g/m^3$, Medium = 110-500 $\mu g/m^3$, High = 500-1000 $\mu g/m^3$

B: The lowest and highest mean value obtained for desiccant materials tested.

Table 12-4. Removal Efficiency Measured from Type II Experiments.

fore, the surface area of the desiccant is an important factor. Higher the surface area of the adsorbent, the higher its adsorption capacity is for all compounds. However, the surface area has to be available in a particular pore size within the adsorbent. At low partial pressures (concentrations), the surface area in the smallest pores which the adsorbate can enter into is the most efficient.

The thermal regeneration capabilities of both desiccants were evaluated qualitatively rather than quantitatively. Two distinct trends were observed: higher regeneration temperature favored the desorption capabilities of both desiccants, and lower inlet contaminant concentration increased the desorption capabilities of the desiccant tested. Further research is needed to determine the quantitive effect of thermal regeneration of desiccants to estimate their potential life expectancy.

References

1. Moschandreas et. al. Control Experiments for the Removal of Selected Pollutants from the Indoor Environment. Topical Report. GRI Contract #5081-251-0575, 1983.

2. Brunauer, S. "The Adsorption of Gases and Vapors." Oxford University Press, London, 1953.

3. Kovach, J.L. "Gas-Phase Adsorption and Air Purification." *Air Pollution Handbook*, Academic Press, 1968.

4. Wright, G.T. "Application for Synthetic Molecular Sieve Zeolites and Silica Gel Towards the Separation for Sulfur Dioxide from Combustion Gases." Virginia Polytechnic Institute, 1979.

5. Vinnikov, L.I.; Mukhlenov, I.P.; Lesokhin, I.G.; and Buzanoua, G.V. "Adsorption of Sulfur Dioxide from Wet Air by Industrial Adsorbents." The Soviet Chemical Industry, No. 2, EEB, 1973.

6. Macriss, R.A.; Rush, W.F.; and Well, S.A. "Air Cleaning Adsorption Process." U.S. Patent 4,012,206, March 15, 1977.

7. Gidaspow, D.; Hariri, H.; and Arastoopour, H. "NO_x Removal with High Capacity Metal Oxides in the Presence of Oxygen." Land EC Process Design and Development, Vol. 20, April, 1981.

8. Gustafson, P. and Smith Jr., S.H. "Removal of Organic Contaminants from Air by Type 13X Molecular Sieve." NRL Report 5560, Dec. 6, 1960.

9. Relwani S.M. et. al. Indoor Air Quality Control Capabilities of Desiccant Materials. Proceedings of the 4th International Conference on Indoor Air Quality and Climate, Vol. 3, pp. 236-240, Aug., 1987.

PART 3

PREVENTIVE INDOOR AIR QUALITY MEASURES

13. SYSTEM DESIGN AND MAINTENANCE GUIDELINES

Peter W. H. Binnie
Vice-President, ACVA Atlantic, Inc.
Fairfax, Virginia

Page

13.1 Introduction

In this chapter, system design guidelines, installation of equipment, and maintenance guidelines will be explored with respect to preventive measures. Among the system design guidelines are ventilation, relative humidity, filtration, and air-handling equipment. The installation measures will include choices of equipment, installation checks, and commissioning. The maintenance guidelines, in addition to pinpointing the areas where regular maintenance and cleaning are vital to indoor air quality (IAQ), will include a proactive monitoring program (PMP).

13.2 Design Guidelines

13.2.1 Ventilation

The architects and engineers who design buildings and their heating, ventilating, and air-conditioning (HVAC) systems depend on recommendations from organizations such as the American Society of Heating, Refrigerating, and Air-Conditioning Engineers (ASHRAE) and the Building Officials Code Administration International, Inc. (BOCA) on ventilation rates. Most states have statutes or building codes which require the builders and developers to incorporate the appropriate mechanisms and tolerances for adequate amounts of outside air to be introduced into the buildings. At the present time, however, there is no legislation enforcing these measures.

ASHRAE Standard 62-1989 (1) for commercial buildings recommends a minimum of 15 cfm/person of fresh air be brought into the building with a preferred rate of 20 cfm/person. According to the BOCA in 1986 (2), this rate should be 20 cfm/person with a preferred rate of 35 cfm/person for certain areas such as conference rooms. These rates will ensure the levels of carbon dioxide remain below 1000 ppm and adequate dilution of any other airborne indoor air contaminants. Naturally, this recommendation assumes that all other recommended pollution control methods are followed. These methods include direct exhausting to the exterior from toilets and kitchens, and where it is known that certain processes will be carried out in buildings such as printing, photocopying and laboratory work involving fume hoods, that the necessary direct extraction and exhaust methods will be affected at the source of the pollution.

In buildings, where designated smoking areas are to be provided, great care must be taken to ensure that the environmental tobacco smoke (ETS) is not concentrated and distributed throughout a floor or the entire building. For example, this can happen if the building system uses a ceiling void return plenum and, to encourage the smoke movement out of the designated smoking area a booster fan is fitted in its false ceiling. This will simply suck the smoke into the ceiling void where it will either be entrained in the return airflow for the floor or building and thus be distributed with the circulating air or, because of the increased positive pressure from the booster fan, the void may become positively pressurized (instead of negatively), resulting in the air/smoke mixture to directly enter the areas adjacent to the smoking area from the void.

A correct design of a designated smoking area will usually require a dedicated exhaust fan and duct system eliminating the air from that area directly to the exterior of the building. Auxiliary "smoke-eating" equipment may also be installed. Such equipment, if employed, will require proper maintenance for continued effective operation.

Assuming that the necessary amount of fresh air is to be brought into the building, the system must be designed to provide the necessary supply airflow rate with which to carry the fresh air and distribute it to the occupied spaces, along with the other requirements of the supply air such as temperature, relative humidity, and cleanness. The cleanness of supply air introduced into a building is a function of the location of the fresh air intakes. Therefore, fresh air intakes must be carefully positioned with respect to prevailing wind directions and patterns caused by adjacent structures and the proposed location of exhaust vents from the building. Fresh air intakes are frequently found adjacent to toilet and kitchen exhausts, restaurant and other trash dumpsites, cooling tower spray drifts, toxic gas vents from hospital sterilizing rooms and laboratories, underground parking garages, and busy streets. Furthermore, fresh air intakes should be designed to minimize the entry of snow and rainwater, and sloped toward the outside to allow water to completely drain away. The intakes should be louvered, wired or protected in a manner to prevent the entry of birds, large insects, small animals, and wind-blown debris.

13.2.2 Relative Humidity

The recommended relative humidity in commercial buildings ranges from approximately 30% to 70%. Below 20% to 25% relative humidity such as in the heating season, the air becomes very dry as well as the uncomfortable buildup and release of static electrical charges. The mucous membranes of the eyes, nose, and throat begin to dry out. This makes them more sensitive to particulate irritation and more susceptible to microbial infections. People wearing contact lenses, asthmatics and those with allergies are particularly prone to this type of drying out condition. Therefore, in designing HVAC systems, it is often necessary to include some degree of humidification provided the building has a sufficiently effective vapor barrier to prevent other construction problems due to condensation water. The choice of a humidification method has important bearings on the microbial content of the air. The safest method to humidify the supply air is the direct injection of steam. In this method, care must be taken to ensure that the source of steam does not contain toxic amines often present in main plant boiler steam as water additives against corrosion, and that precautions are taken to ensure water leaks from the equipment can be completely drained away.

If cold-water type humidifiers are used, the water should be from a potable or other treated source and ideally the excess water should be drained away and not recirculated. In any case, these types of units must be kept extremely clean and free of microbial slime by regular draining and cleaning.

Spray-type humidifiers are now notorious for their ability to provide a source of moisture and food for microorganisms and along with their intermittent use depending on the seasons are ideal disseminators of all types of infective and allergenic materials into the supply airstream. They should be removed from the existing systems, where possible.

The use of individual portable "cold mist" type vaporizers should be discouraged as they readily become contaminated with microorganisms which are easily disseminated into the adjacent air as aerosols. The range of effectiveness of these vaporizers is so small that it is unlikely they have any real effect on relative humidity.

When the relative humidity levels rise above 70%, in humid summer months of certain climatic zones, the air becomes uncomfortably sticky for the building occupants and conditions become more favorable for germination of the spores of many molds and fungi. Under these conditions when outside air is introduced into a building for ventilation, it needs to be dehumidified, usually by cooling. This may mean that reheat coils will have to be used to bring the air temperature back to an acceptable comfort level for the occupants.

13.2.3 Filtration

Through surveys, inadequate ventilation and faulty filtration are the two most common factors found in commercial buildings (3, 4).

The filters installed, even in new buildings, are mostly simple furnace-type filters which consist of one layer of glass fiber material which will stop moths and butterflies. Building studies frequently find:

- Missing or improperly installed filters allowing the air to take the least line of resistance and avoid the filters;

- Filters damaged or extremely overloaded with dirt that they buckle under pressure;

- Filters excessively loaded with dirt on which microbes grow and be carried in the supply air from their downside;

- Torn or otherwise damaged bag filters;

- Automatic roll-on filters that do not roll on or completely rolled out;

- Filters wetted by rain or snow entering the system or from carry over from chill or spray coils, making them more susceptible to microbial growth;

- Oiled-filter systems, which depend on the oiling process for their effectiveness, malfunctioning; and

- Electrostatic filters that are not operational because of electrical faults or excessively dirty electrodes.

Originally, filters were designed to be installed upstream of the HVAC system equipment to protect it from being clogged with dust and dirt. It was not thought necessary to consider that the people breathing the air might benefit from a reduction in particulate content. This tradition is firmly embedded with designers and only in some hospitals or other buildings where extra clean air is required are filters installed on the down side of plant or, even as should be the case with hospital operating rooms, at the last point possible in the ducts before the air is supplied to the receiving area. The air-conditioning equipment itself generates a considerable amount of particulate matter, both viable and nonviable particularly if it is not kept clean, and this is carried along with the supply air to occupied areas if filters are installed only at the front of the air-handling units.

An ideal configuration for a commercial building involves a set of prefilters on the mixture of return and fresh air (to protect the equipment) and then a set of medium-efficiency filters downstream of the unit. The prefilter can be a one- or two-inch polymedia type and the downstream filters can be two-inch pleated panel or similar extended surface filters, such as bag filters. It would also extend the life of the downstream filters if a set of polymedia filters were fitted in front of them. In practice, we will be satisfied with a well-fitted, good quality, medium-efficiency pleated panel or bag filters protected by low-efficiency polymedia frame or roll-on filters. Many surveys have shown this arrangement to provide satisfactory IAQ provided there is a satisfactory overall equipment maintenance and cleaning program in force.

Filter technology has improved immensely over the last few years and it is now possible to manufacture filters with a high level of particle retention without a dramatic loss in airflow provided that they are properly installed and maintained. A common way of determining when filters are due for change is to measure the pressure drop across the filter bank by a permanently installed manometer. In addition to this however, visual inspections should also be made of the filters to ensure that they are in position, not damaged, wet or misplaced, or overgrown with fungi.

Electrostatic precipitation filters are very efficient when they are clean and operate properly having an efficiency of 90% to 95% at the size of 0.5-micron diameter. Their disadvantages are due to their high initial cost; if a power failure occurs, their efficiency drops to zero and they require intensive maintenance and cleaning to keep them operating at full efficiency. If installed ideally, they should also have low- to medium-efficiency filters fitted upstream and downstream to help keep the electrodes clean and to trap the ash they produce as well as acting as a back-up should power fail. Older models of electrostatic precipitators have a problem with ozone production not present in newer models unless the unit is left on with no air being drawn across the plates. Where gaseous contaminates are present, in outside air such as in heavy industrial areas, it may be necessary to clean the air before it can be used as supply air in a building. This may be accomplished by using special filtration methods such as activated charcoal, and where odors may be a problem they may be oxidized by passing the air through an agent such as alumina impregnated with an oxidizing agent such as potassium permanganate. In buildings such as museums and art galleries, these steps may be taken where such contamination may have deleterious effects on the artifacts.

13.2.4 Air-Handling Equipment

The majority of equipment design, in building surveys, indicate that the designer has not given much consideration to the fact that the equipment needs to be inspected, cleaned and repaired. Especially, small air-handling units do not have easy access to their chambers and it is rare to find adequate access to long duct-works installed in many large buildings. Adequate design practices in the development of systems and equipment are those which will produce the necessary volumes of clean, conditioned air and at the same time not allow dust and direct accumulations and free water to develop within these systems which can contaminate the air. The recommendations include:

- The specifications of materials suitable for the job. Often chill coils are mounted in a framework of thinly galvanized metal that is constantly exposed to water rusts long before the useful life of the coils is ended. Condensate trays and other parts of air-handling units to be exposed to water should be manufactured from a noncorrosive metal or treated to resist corrosion.

- Chill coils and spray-type humidifiers also rust and corrode if they are not made from stainless steel nor protected. In the dry season, the particulate matter consisting of rust, water salts, and corrosion products are carried in the airstream to the occupied area. Additionally, any bacterial, fungous, protozoan, or slime growth will die down as the water evaporates and will dry out. Many will sporulate and many will become desiccated and be carried in the supply air to cause allergic reactions in susceptible people in the receiving areas; even some fragments of dead cells can be strongly antigenic. When water returns to the dried out trays, any spores left will germinate to give new growth and the cycle begins again. Also rust and corrosion inevitably lead to leaks, which means water may reach other parts of the system or mechanical room where no drains are present to allow further microbial growth to occur.

- Access doors should be fitted to every chamber in air-handling units and easy access must be available for filters to be inspected and replaced.

- Steam humidification should be used with proper provision of drains to prevent water from leaks collecting in ducts.

- Condensate trays under chill coils must be wide enough to collect from the complete width of the coils and be properly sloped and drained; the drain should have a trap strong enough to withstand the negative pressure produced by the fan if that is located downstream.

- Baffles should be fitted where necessary to prevent water carry-over from chill coils into downstream chambers and should be resistant to rust and corrosion.

- Insulation of air-handling units should either be done externally or between double walls of chambers. Internal glass fiber insulation especially in fan chambers and main air supply ducts should be avoided. Internal insulation acts as a dust and dirt trap and, when it is wet, becomes an ideal breeding ground for microbes. In addition, the surface seal on this type of insulation is seldom good enough to withstand the constant air beating for long with the result that tears develop which spill the glass fibers into the airstream and provide another pollutant.

- Branch ducts should be made from galvanized steel with external insulation and be provided with access doors conveniently located for inspection of reheat coils, variable-air-volume (VAV) boxes, etc., and at places where moisture may collect.

- Air supply diffusers and return air grilles should not be located in the receiving areas close to each other to cause shortcircuiting. This may lead to stratification of the air with reduced circulation so that people seated at desks, for instance, may not actually receive fresh air.

- Ceiling voids should not be used as supply air plenums. They are naturally dusty and dirty and air passing through them will invariably becomes contaminated. A common faulty design feature is one where the air is supplied to the ceiling void and then expected to find its way through slots or small holes on the ceiling tiles to the rooms below. In a very short time, these small passage ways become blocked with dirt and the air pushes through cracks and joints revealing dirt streaks on the ceiling.

- It is a bad practice to design a system which involves VAV boxes or fan coil units drawing air from ceiling voids under certain conditions as a source of recirculated air.

- The introduction of free fresh air into a ceiling void plenum where this is the source of other air supply equipment to draw from. This is another design which does not work.

- Under satisfied temperature conditions, VAV or mixing boxes in a ceiling void may actually cutoff the supply air to an occupied area when it becomes heavily occupied with people due to additional heat generation, thereby, reducing the amount of supply air coming in when it is most needed to dilute carbon dioxide and other indoor air pollutants.

- Ancillary air supply equipment such as perimeter fan coil units or induction units should be accessible for inspection and cleaning, and should be equipped with drains if condensate will collect and be provided with adequate filters.

13.3 Installation

13.3.1 Choice of Equipment

A business, before purchasing and installation, should seek outside assistance from organizations who will advise without bias on the effectiveness of equipment and design of HVAC equipment in accordance with the submitted specifications. Obviously, these specifications should then be adhered to and under no circumstances should inferior quality equipment be substituted. Care should be taken in the selection of suppliers and installers.

13.3.2 Installation Checks

Special attention must be paid to the details as air-handling units and their accompanying air supply and return systems are put together. Although some of the recommendations below appear to be trivial, failure to follow them may lead to failure of the entire system.

- Filter banks must be installed so that no air leaks exist which can lead to air bypassing them.

- Condensate trays must be checked to ensure that there is sufficient slope to allow complete water run-off.

- Drain pipes from condensate trays should be attached flush to the tray to ensure complete draining.

- Drains must be checked to ensure they are adequate to cope with the expected volumes of water.

- The depths of drain traps should be checked to ensure that they can cope with any negative pressure generated by fans.

- Ductwork provision should be made for the removal of moisture adjacent to the humidifiers and coils.

- Dampers in ducts should be checked for proper functioning and fitting.

- Where internal fibrous glass insulation is used, all joints must be properly sealed and the surface seal should not be damaged.

13.3.3 Commissioning

It is common practice for a newly installed HVAC system to be commissioned before it is finally accepted by the client to ensure that it is functioning in accordance with the specifications. A detailed discussion on commissioning is reserved for Chapter 14, Part 3.

13.4 Maintenance Guidelines

Regular maintenance plays a chief role in preserving system efficiency. Regardless of how expensive the equipment and how carefully the installation is monitored, the system will start to deteriorate literally from the first day on. The rate of deterioration of a system is a function of its quality and frequency of maintenance and cleaning. In commercial office buildings, visible areas are constantly monitored so as to reduce the amount of dust and dirt because it is unhygienic, unpleasant, and uncomfortable if these areas are dirty. On the other hand, the unseen places where such buildup of dust and dirt occurs, are more likely to be forgotten or ignored.

13.4.1 Areas of Concern

Some areas where regular maintenance and cleaning are vital to IAQ are as follows:

- Fresh or outside air intakes should be checked regularly to ensure that they are not obstructed; their grilles are intact and clear; and their dampers are in good condition, operating properly and set to allow a minimum of 20 cfm/person of outside air at all times of occupancy.

- Filters should be checked and regularly maintained to ensure that they function properly to remove airborne particulates from the supply air rather than contributing to pollution. Visual inspections as well as monitoring manometers or other filter changing indicators should be provided so that filter fit can be checked as well as damage noted, and that roll-on filters are rolling on and are not rolled out. Filters should be changed as soon as they are due with the unit turned off and operators wearing protective masks. The discarded filters should be placed and sealed in plastic bags upon removal and the new filters should be inserted immediately. Filters should not be allowed to become wet as this will allow microbes to grow and may also release toxic vapors associated with bonding materials used in some filters into the supply airstream.

- Heater coils in air-handling units do not normally cause problems unless they are in a position to become splashed by carry-over from chill coils or spray-type humidifiers when they rust very quickly because of high temperatures.

- Chill coils and spray-type humidifiers are the main sources of free water and, therefore, potential microbial problems in air-handling units. They can both contain stagnant water which becomes contaminated with microorganisms and then releases spores and aerosols into the supply air. Cooling coils should be operated at sufficiently low temperatures to allow adequate dehumidification of the supply air to the occupied spaces.

  Chill coil condensate trays must be kept free of dirt and microbial slime and well drained. Slow release chemical sachets will help keep microbial growth in check and drains clear. Spray-type humidifiers with spray nozzles and jets and a recirculating water reservoir should be eliminated from air systems if possible; if not, they should be drained and cleaned out when not in use and subjected to rigorous and regular cleaning and slime removal when in use. Water may also enter otherwise dry air-handling unit chambers from leaks in chill or heat coils. This is a particularly dangerous situation.

Ideally, all systems should be made from stainless steel or treated with a protective anticorrosion cover. The key to prevention of these types of problems is regular maintenance and inspection with prompt action to repair leaks and blockages and removal of free water from the system.

- Cooling towers should be constructed from corrosion-resistant materials and designed for easy access for maintenance and cleaning (especially for fill and sump pumps). All internal corners and edges should be rounded to facilitate cleaning, and porous materials should not be used for surfaces which will be wetted. Cooling towers should be installed with suitable water treatment and bleed-off systems built-in to help control corrosion, scale and algal growth. A regular cleaning and maintenance program using appropriate biocides known to be effective against legionella species is essential.

- Main and branch ducts should be checked regularly for buildup of dirt, blockages, and air leaks. Damaged internal fibrous glass insulation should be repaired promptly before reheat coils, turning vanes, and dampers become clogged. External insulation should be also checked for damage and repaired to limit the the dispersal of glass fibers into the ceiling void, etc.

- Exhaust and return ducts carry air which is contaminated. Exhaust and return ducts should be checked as they are often blocked with accumulations of textile fiber. These may develop at grilles, turning vanes, and dampers. If the air cannot be extracted from an area, supply air will not easily enter and, reduced ventilation and circulation problems will develop.

- Final distribution equipment should be checked regularly. In many cases, adjustable diffusers control the actual volume of air delivered to receiving areas. If they become clogged, they may upset the balance of the whole system. Revealing dust streaks and stains on ceilings and walls around diffusers may be an indication of dirt buildup in ducts to such an extent that it is spilling out with the supply air.

- Ancillary equipment such as fan coil units and induction units also require regular inspection and cleaning. They may have same types of problems as the full size air-handling units. Many contain chill coils which will produce condensate water, and they may become a source of microbial growth and dissemination if not kept clear and clean.

13.4.2 Proactive Monitoring Program

While visual inspection by maintenance personnel is useful in determining the general condition of an HVAC system, the most reliable way to assess the condition of HVAC systems and the quality of the air supplied by these systems is a regular PMP carried out by a firm specializing in this type of work. The inspections and measurements should be performed by professionals with appropriate instrumentation and experience allowing objective evaluations of the system conditions and quality of the air supplied. If this type of monitoring is utilized on a six monthly basis so that both the heating and cooling cycles are covered, changes in observations and measurements will alert the management to deterioration in system parts or IAQ in certain occupied areas. This will allow remedial action to be taken before IAQ deteriorates to the degree where occupant complaints increase drastically. This type of monitoring will include:

- Carbon dioxide levels;
- Filter retention efficiencies;
- Airborne particle counts and weights;
- Airborne, water, and surface microbial sampling;
- Particulate surface sampling;
- Visual inspection, directly and indirectly with borescope;
- Installation and analysis of continuous air-sampling devices;
- Asbestos bulk analysis and airborne fiber counts, if appropriate; and
- Other sampling and measurements as required.

The results are collated and reviewed. Each test, contributing information on the HVAC system and IAQ will help provide a full picture of the overall conditions. Practical recommendations can then be made to rectify a bad situation or to prevent one from occurring. The PMP provides a six monthly checkup with the purpose of preventing deterioration of IAQ rather than trying to cure an existing IAQ problem.

13.4.3 Other Factors

According to some psychologists, some of the IAQ complaints are partly due to a feeling of losing touch with the outside world. Their inability to open and close windows, not being able to see the natural colors outdoors because of tinted glass, and working under artificial light and having no control of the temperature or airflow around them may be the source of their discomfort. Many employees who feel this way are tied to a desk or area, doing daily repetitive work.

A variation on this theme is described by a Virginia technical institute professor who occasionally receives requests to investigate what he now calls "delusory parasitosis." This involves one or a few members of a group of people who have an unusual fear of insects and can infect the rest of the group with their concern that an infestation is occurring. Medical inspections may even reveal bite marks on skin but thorough investigations fail to identify any potential man-eating insects. Although, in some cases, glass fiber has been identified as the cause of skin itching, no frequently occurring causes are found to be the sources of the problem. In such cases, it is necessary to implement some type of social solution. Without any doubt these mental aspects are a part of the sick-building syndrome (SBS) and warrant investigation.

13.4.4 Removal of Contamination

In cases, where IAQ has deteriorated so badly and the accumulations of dirt in ducts and other parts of the HVAC system are so heavy, one of the recommendations may be to clean the entire or part of the system. Sometimes the rule for cleaning out accumulated dust and dirt is very simple: if an air-handling unit chamber or main duct is knee deep in dirt, one can clean it with a shovel and bag. In other cases because of the constraints placed by the design of the system or its location, other more complicated methods for cleaning have to be developed. A wide range of methods has been used with varying degrees of success such as air-blowing and high-performance vacuuming methods, revolving brushes attached to vacuum lines, air jets on lances, large vacuums with high efficiency particulate air (HEPA) filters attached, and some combinations of the above.

The main goal of the cleaning operation is to remove as much of the buildup of dust and dirt as possible from the internals of the various parts of HVAC equipment in a controlled manner. During this process, disruption of the normal operations of the building should be minimized. Physical removal of dirt may significantly reduce the excessive growth of bacteria and fungi found in chambers or ducts. It may even be reduced further and often to controllable levels using a safe, nontoxic, broad spectrum sanitizing fluid. In cases where there are species present in the HVAC system causing allergic reactions, it is important to minimize them by sanitizing treatments at regular intervals along with good cleaning and maintenance. The frequency of application must be controlled by microbial sampling. There is no one correct method to clean HVAC systems and their ducts and, as no two systems are identical, each one requires the development of on-site techniques to get into the lengths of hidden ducts and clean them out.

13.5 Conclusions

According to some estimates, we spend 80% to 90% of our time indoors. It is certain that this figure will rise in the future as more indoor shopping malls, leisure centers, and superdomes are built. Eventually, the SBS will be tackled by the designers, engineers, and architects at the drawing board stage and every design team will include some type of an environmental worker. His function will be to ensure that each building includes those of HVAC systems and equipment known to be effective in enhancing IAQ.

In the meantime, to overcome the difficulties of inaccessibility in today's buildings and to attack the problems of contaminated indoor air, it has been necessary to devise a new approach. This has been achieved using knowledge, skills, and experience derived from several disciplines and practices by a new breed of investigators for the solution to the problem of indoor air pollution calls for the scientific application of technical skills in biochemistry, microbiology, and epidemiology. The engineer's working knowledge of HVAC systems, the expertise of the building service manager, the design imagination of architecture and interior design, and a knowledge of industrial psychology (to some extent) are among them.

References

1. ASHRAE Standard 62-1989. "Ventilation for Acceptable Indoor Air Quality." Atlanta, GA, 1989.

2. Building Officials Codes Administration International, Inc., Supplement, IL, 1986.

3. Melius, J.; Wallingford, K.M.; Keenleyside, R.; and Carpenter, J. ''Indoor Air Quality - The NIOSH Experience.'' *Ann Am. Conf. Gov. Ind. Hyg.*, 10, p. 4, 1984.

14. INDOOR AIR QUALITY AND COMMISSIONING

Carl N. Lawson
Vice-President, Liebtag, Robinson & Wingfield, Inc.
Tampa, Florida

14.1 Introduction

Two of the most important aspects of designing new and retrofitting old buildings are commissioning and indoor air quality (IAQ). In this chapter, we will discuss why commissioning and IAQ play such an important role in today's new buildings.

Commissioning and IAQ fall into two general categories: (a) new building construction projects and major expansion or renovation projects, and (b) renovation projects with minor or no changes to the basic building structure. Although the methods, intended purposes, and the functions of commissioning and IAQ are identical, they differ only in the level and extent of services provided.

Until now building commissioning that existed in the construction industry for many years, was a simple process that no one really cared about. Now that our buildings have become so sophisticated, commissioning has gained a great importance and, unfortunately, is very difficult to accomplish. Commissioning is well understood in the process and production industries, where a new plant is brought on-line through of partial performance tests until it is able to attain and sustain full output. Major petrochemical and process engineering firms have their own specialists in the field who often travel internationally to fulfill their responsibilities.

Today, the manner in which new buildings are completed and turned over to the owners has become a very complex issue. The same thing is true for IAQ. Although IAQ has been around for centuries, no one ever gave it much thought until recently when people started to complain about their workplaces (1). Such complaints include eye, ear and throat irritation, headaches, vomiting, and unpleasant odors. The owners and managers faced with the development and leasing of new spaces in today's demanding market must address the issue of IAQ. Market research by developers and corporate tenants shows that the protection of tenants and productivity at the workplace are as important as the location, if not more. For example, the cost for a tenant to staff and occupy a typical office building in a prime location is $250 per square foot per year. Of this, $248 is for salaries, benefits, training, and rent. The remaining two dollars or less is for providing the required environmental services. If this small allocation is not distributed properly, or the systems are not designed to function properly and maintained poorly, or if energy efficiency measures result in cutbacks on this small allocation, the building may become a sick building and health of the occupants and productivity at the workplace may be at great risk.

A recent article in the journal of the American Institute of Architecture (2) warns that the single most important area of liability litigation facing building designers is the environmental performance of buildings. For example, recent areas of litigation have included asbestos materials, formaldehyde products, and microbial growth in mechanical and plumbing systems. According to the estimates, 90% of the currently available office buildings has the potential to have such problems. Fortunately, these problems can be eliminated. To avoid sick buildings, the architects and engineers must understand the nature of health and comfort problems resulting from poor building design. Healthy and energy-efficient buildings do not need to be designed in a fashion or at a cost that detracts from their appeal to conventional developers and office tenants. Carefully applied existing technology and proper environmental site planning can lead to a building design satisfying energy efficiency and IAQ without compromising the conventional office space marketing requirements. In fact, with today's office leasing markets sensitive to IAQ and energy issues, capital cost premiums associated with environmental or energy improvements may be recaptured during the term of the lease if these improvements are intelligently presented to the owners and tenants.

14.2 The Need For Commissioning (3)

Commissioning is defined as the process of verifying and documenting the performance of systems to ensure their operation in accordance with the design intent. Design intent includes the design criteria and assumptions and the description of the systems including the intended operation and performance in all seasonal modes at part- and full-loads, with all key assumptions and compromises made. As one reviews a set of plans and specifications intended for construction, it is often impossible to determine the design intent. It is certainly impossible to determine assumptions made in load calculations. Unless explicitly stated, it is impossible to determine which equipment, if any, is redundant in each seasonal mode of operation.

There is a need for the design consultant to document his design intent. This documentation should be available to the owners, contractors, commissioning authority, and especially the operating personnel so that the system can be commissioned and operated as intended.

The design criteria should include the following as a minimum:

- Indoor design conditions for all seasons;
- Outdoor design conditions for all seasons;
- Assumed density of building population and hours of use;
- Assumed electrical load for lighting and power use;
- Ventilation design assumption. They are verification of floor and wall covering, other out-gassing materials in the original design or added at occupancy;
- Special loads, computer rooms, printing, special photocopiers, etc.;
- Definition of building envelope, including the types and characteristics of materials and assumed infiltration;
- Code requirements;
- IAQ design criteria;
- Noise criteria;
- Fire and life safety requirements; and
- Energy efficiency and projected operating costs.

The description of the heating, ventilating and air-conditioning (HVAC) systems should include the following as a minimum:

- Basic system types;
- Major components;
- Capacity and sizing criteria;
- Redundancy provisions;
- Intended operation in each seasonal mode, including designed changeover conditions;
- Changeover procedures;
- Part-load operational strategies for each season;

- Occupied/unoccupied modes of operation for each season;
- Design setpoints for control system, including permissible limits of adjustments;
- Operation of system components in life safety modes;
- Energy conservation procedures; and
- Other engineered operating modes of the system.

Every building is unique. The design consultant must consider each of the above when designing the system. Many buildings are adjusted, commissioned, and operated by people who have no idea of the design intent. When the commissioning is left to technicians who do not understand the system, we risk compromising the design. There is a need for the commissioning process to be an integral as part of the project specification. This specification must include the qualifications of the commissioning authority, a person in charge of directing the commissioning process. The commissioning authority may be the designer or installing contractor or a third-party specifically involved in the commissioning process; this person must be capable of fully understanding the design criteria and the system description and operating modes (4). The specification must then define logical and sequential procedures for various commissioning activities from individual components to subsystems to total system performance in all modes of operation for full- and part-loads in all seasonal conditions. The commissioning process should be documented and this documentation be reviewed by the designer to ensure conformance with the design intent. The extent of the required documentation must be a part of the commissioning specification.

Probably the most important part of commissioning a system is the proper commissioning of the building operators. The building operators must understand the concepts of the design, the proper operation of the system in each mode, the permissible adjustments that can be made in each operational mode without any compromise, diagnosis of problems in performance, and the ways to correct them. System operation is much more than simply maintaining equipment.

The commissioning authority must direct the training program for the building operators using the design criteria, systems description, and the commissioning documentation as the basis of instruction. Also, various vendors, manufacturers, and subcontractors must participate in the training program under the direction of the commissioning authority.

14.3 The Commissioning Team and Its Role (4,5)

14.3.1 Responsibility of Owner

The owners have three main responsibilities with respect to commissioning. First, they should define the responsibilities the architects and consulting engineers are expected to undertake. This permits them to determine appropriate fees for the commissioning-related work and include them in their proposals. Four specific items are identified as follows:

(a) Ensure that building systems functional performance verification tests will be carried out by the contractor. This is to be described in a properly detailed commissioning specification included in the contract documents.

(b) Verify and document the results of these functional performance tests. This will require the on-site presence of the architect/consulting engineer to observe all such tests.

(c) Provide proper instructions to the operating personnel with respect to design intent, operating characteristics, and capabilities and limitations of the HVAC systems. This will normally be done in a conference setting.

(d) Provide a brief written outline of the foregoing instructional material to be included in the Operating Manual. It is assumed that the designer is in a much better position to provide such material than the contractor.

Secondly, Operation and Maintenance should assign an operating staff early in the design/construction phase to allow its involvement in the commissioning process. However, this does not mean that a permanent, full-time operating staff is to be assigned to a building long before it is completed. In fact, most buildings do not have full-time operators, even after completion because they are too small. However, it is important to establish operating responsibilities at this time. The specific activities the operating staff participate in are:

(a) Contribution to design proposals from an operations and maintenance point of view. Such involvement would be more extensive on major or complex projects in comparison to small projects.

(b) Periodic on-site availability during construction to become familiar with the installation and receive necessary instructions. This would include the formal instructional session(s) by the consulting engineer and operational demonstrations/explanations by the contractor.

(c) Observation of some or all of the systems functional performance verification tests carried out by the contractor.

Finally, the owner should provide professional expertise to support the commissioning process (5). This will be particularly important during the initial planning stages of the project when the consulting engineers are still uncertain of the expectations and the owners are still feeling their way to some degree. This will involve:

(a) An overview of the contract documents to ensure that the commissioning requirements in the contract documents are satisfactory.

(b) Monitoring preparations for the systems functional performance tests and ensuring that they are carried out properly.

(c) Identification of indoor air pollution sources and contaminants (1).

14.3.2 The Role of Architects and Engineers in Commissioning (2)

The most effective approach to having a healthy building is to design and build one from the outset. To accomplish this, the architects, engineers, and their clients must employ a preventive approach called commissioning. The term commissioning refers to a comprehensive evaluation of a building project to guarantee the effective performance of the IAQ control systems. Commissioning defines IAQ as an essential design objective.

The architects and engineering consultants must work as a team and actively pursue their roles in commissioning and IAQ. Modern buildings encompass many system control elements such as temperature, humidity, lighting, noise level, the intended physical and architectural characteristics of the space, the intended sense of enclosure, and the types and proximity of occupants. IAQ is a result of the interaction of all these factors.

Traditionally, the architects and engineers have not been involved in commissioning and IAQ except secondarily in the role of the prime consultant. The prime consultant has been responsible only for the coordination of the work of the secondary consultants and, upon completion of the building, as a responsible person to whom system failures are reported to commence the remedial process.

In the traditional design process, the architects and engineers have defined the environmental control systems and building components. After construction and system installation, inspectors and contractors have evaluated the effectiveness of systems through fairly narrow guidelines. The building inspector reports back to the contractor, who then reports back to the engineer, who then reports to the architect, who then certifies the building as substantially complete when all inspections and approvals have been given.

The architects and engineers must now take a more active leadership in commissioning and IAQ. The prime consultant is the only person who has the overall picture of what a building and its systems need to do. The architects and engineers need to know what to achieve, what their client will accept and, in the long run, that the building they have designed will meet the owner's requirements. In achieving these goals, the architects and engineers must recognize that:

- Meeting design requirements of codes does not guarantee an effective IAQ control system,

- Air balancing and certified test reports do not necessarily imply that a building functions as effective or environmentally as designed, and

- Ventilation strategies for buildings must be verified again two to three years down the road.

The architects and engineers should involve in commissioning from the predesign phase of a project through the completion phase and beyond to ensure the on-going performance of the building. The latter involves the most significant change to what is traditionally perceived as the role of an architect/engineer in building design and construction. When an architect and engineer undertake a building project, they, along with their firm implicitly accept responsibility. To justify their responsibility, the architect and engineer should be awarded the position of commissioning authority for the continued coordination of the building evaluation; but the overall commissioning should be by a third-party to avoid any conflict of interest.

14.3.3 Responsibility of Contractor (4)

The contractor must plan and carry out the physical activities required in the commissioning process. The secret to success is an effective commissioning plan. When a commissioning plan is developed early in the project, the contractor is tuned in to the requirements. When one is not developed or the initial efforts are inadequate, one can expect difficulties in achieving satisfactory completion. There are several very important areas that must be accomplished by the contractor.

14.3.3.1 Personnel Selection. Select a team leader who is capable of starting and understanding the mechanical equipment used. He must have a personality that will allow him to coordinate work with all trades and the professionals involved.

14.3.3.2 Scheduling a Pre-Commissioning Meeting. This meeting should be attended by the commissioning authority, design engineer, mechanical start-up supervisor in charge of the project, control subcontractor, testing and balancing contractor, sheet metal superintendent, piping superintendent, and the owner. At this meeting, the function of the mechanical systems and intent of operation are addressed. Questions concerning the types of systems, equipment, special features, and control sequence should be answered. Potential problems that might exist should be thoroughly explored prior to start-up.

14.3.3.3 System Start-Up. One of the greatest difficulties standing in the way of smooth commissioning process is the necessity (usually caused by work conditions) to start the system in segments. Once a building is enclosed in glass or masonry skin, some type of reasonable environment must be provided for those working in the enclosure or productivity will decline. This causes the system to be started in segments. As a result, there exists problems with equipment warranty and loss of continuity for the start-up team. If proper records are kept, the start-up team can continue where they left off when they return for the final start-up. Assuming that the partial start-up is accomplished, we will now tackle the actual complete start-up in the order as follows.

14.3.3.3.1 Preparation of Start-Up

(a) Coordination with Electrical Contractor. Since most of the mechanical equipment have electrical connections, verify to ensure all connections, controls, safeties, and interlocking are correct.

(b) Sheet Metal Coordination. Verify that all duct connections are made and all dampers are installed and complete in time to start all air systems.

(c) Piping Coordination. Verify that all piping connections are made and control valves installed, and the system has been properly flushed and refilled. Establish a specific time to start-up pump systems.

(d) Control System Coordination. Verify that all systems are complete and ready to operate. The system must be checked once more in all operating sequences.

(e) Coordination with Manufacturers About Equipment Start-Up Responsibility. Chillers, multispeed pumping systems, cooling towers, air compressors, chemical treatment, and boiler systems must be started and given an operational run by their manufacturers. These runs should be documented in written reports for the purpose of verification.

(f) Testing and Balancing Contractor Coordination. The testing and balancing contractor should be notified of the schedule for the fans, pumps and chillers, and the approximate time for a full system performance test.

(g) Facility Automation System. The facility automation system is a newcomer and causes more problems at this time than any other phase of the mechanical system start-up. It must be coordinated with the mechanical system since it often controls all segments of the system. For this reason, the personnel conducting the mechanical start-up must coordinate this with the contractor to determine what steps must be taken to bypass segments of the automation system to start the system.

14.3.3.3.2 Actual Start-Up. Verify the following:

(a) Alignment of all direct coupled equipment, such as fans and pumps. This is an important step and caution must be exercised to meet all factory tolerances.

(b) Check all blower and motor belts to ensure that belts are properly aligned and that there is proper tension in the belts. Make sure the belt guards are not lost since they must be installed after final adjustments.

(c) Lubricate all moving parts and check off all equipment to be lubricated against a checklist. Too often a bearing fails within 90 days due to improper lubrication. Remove all tie-downs on fans and compressors. Check rotation of all electrically driven equipment.

(d) Sequence all controls and interlocks. This is a dry run without load connections made. Ensure that all sequences are in accordance with the specified sequence.

(e) Ensure that all dampers are in the proper position for air units prior to start-up and that they travel freely. Inspect overloads on all starters for proper setting.

(f) Start all units, including fans, air-handling units and pumps.

(g) Start the chiller and all heating components.

(h) Use an ammeter to check all components listed above three items to ensure that they are within their rated tolerances.

(i) Verify that testing and balancing have been started, and check to ensure that estimated completion date will be met.

(j) Make corrections required for proper system operation.

14.3.3.3.3 Final Start-Up - Complete Performance Inspection. Now that all equipment have been started, it is time to check off each item involved in the total mechanical process in order that the complete HVAC system can be tested as an integrated system. To accomplish this, the start-up personnel should proceed with the following:

(a) Temperature Control System. Check sequence of each individual zone and each individual air unit to ensure that they are operating as specified. Place the main air systems in operation in full automatic control and inspect the performance of the system sequence. As these units and systems are inspected, documentation should be kept so that discrepancies and satisfactory completion of the test can be noted. All final temperature setpoints for the different points of control should be documented.

(b) Facility Automation System. It is necessary to test the performance of this system separately, and the same sequence should be performed as with the temperature control system. Each point at which the system starts, stops, or controls must be inspected and the performance be verified in the same manner as the temperature control system.

(c) Document Test Results. Monitoring, report capability and other requirements for the system not affecting the actual operation of the mechanical system should be inspected and documented at each point. All discrepancies should be noted and corrected in the same manner as the temperature control system.

(d) Testing and Balancing. All testing and balancing reports should be turned in to the project manager. These reports should be carefully reviewed at a meeting consisting of the commissioning authority, project engineer, start-up personnel, and the test balancing supervisor. Potential problem areas should be reported at this meeting. Final reports should be turned over to the project manager following all corrections and final adjustments.

(e) Equipment Documentation. The start-up personnel has the task of documenting each piece of equipment furnished by the mechanical contractor. This entails the listing of the model, serial number, and the manufacturer. This list is then turned over to the project manager and incorporated in the operation and maintenance manual that will be supplied to the owner.

14.3.3.4 Final Commissioning. After final start-up, a commissioning meeting should be held with representatives of the mechanical contractor, commissioning authority, consulting engineer, owner's representative, and the operation and maintenance personnel. All final documents to be submitted to the owner are reviewed at this meeting. These include: (a) as-built mechanical drawings, (b) as-built temperature control drawings, (c) as-built automation drawings and documentation, (d) instruction manuals, (e) testing and balancing reports, (f) maintenance and repair manuals, (g) building inspector's approval, (h) state certificates, (i) equipment manufacture start-up, and (j) warranties/guarantees.

In addition, a general discussion of the system should cover all questions about sequences, setpoints, and operation. Following this meeting, all documents listed above should be turned over to the general contractor or construction manager to be transmitted to the proper authorities. At this point, a date for owner training should be set. The number of training hours necessary, if not specified, should be determined along with the number of sessions required to meet this goal. As previously recommended, this segment can be shortened by the involvement of the operating personnel.

14.4 Commissioning Plan (6)

The HVAC commissioning plan should detail the implementation of the commissioning process. It should include the requirements that each party involved in the commissioning process will have to accomplish, including sequence, scheduling, documentation requirements, verification procedures, staffing requirements, etc. The contract documents should describe the content of the plan and the coordination, submission, and approval of the proposed commissioning process. The plan should outline the following:

- Responsibility of each trade affected by HVAC commissioning, as required by appropriate sections of the specification.

- Requirements for documentation as listed elsewhere.

- Requirements for documentation of the HVAC tests and inspections required by code authorities.

- Requirements for the HVAC commissioning plan during the specified operational seasons at part- and full-loads.

- Format of the training program for operation and maintenance personnel.

The contract documents should refer to other sections in order to provide the necessary coordination and cooperation of other trades in the preparation and implementation of the commissioning plan. Documentation, such as operation and maintenance manuals, test and balance report warranties, inspections, tests, etc., should be assembled and provided to the owner. The format of these data should be outlined.

The extent of the commissioning plan should be outlined, such as incremental load of HVAC systems, seasonal operation, and post-construction verification. The extent of owner personnel training should be also outlined and the format of the training program, slides, manuals, and videotape presentations should be described. Any post-construction training required should be specified.

14.5 Verification Procedure (7)

The designer should describe the tests and demonstrations to be performed during the commissioning phase of the construction. As presented in the ASHRAE Guideline 1P, the description should include the work to be performed by the construction team and the documentation procedures to be followed. The specification prepared by the designer should include the inspection and verification that individual components and systems are set and ready to operate, that preliminary testing and balancing are complete with verification, and the extent of other functional performance testing to be performed.

The sample specification (7) shown in the ASHRAE Guideline 1P appendix reflects the current industry specification practice format. The resulting commissioning specification is more performance-oriented and is similar in presentation to testing and balancing specifications that refer to recognized organizational standards as opposed to a prescriptive specification. The designer should develop a description or checklist of commissioning activities for each component, subsystems, and total building HVAC system. This checklist, once developed, is transferable to future work but it should be revised for each specific job and allow confirmation that the design and intended performance criteria are demonstrated.

14.6 Training of Personnel

The final step of the commissioning process is the personnel training. At this point, all the documentation should have been forwarded to the owner, and the general contractor should have set the dates with the owner. Training sessions should be outlined with dates and time allotments for each session. These dates must be setup to the satisfaction of all concerned. The original format as developed by the consulting engineer and mechanical contractor should be followed:

(a) System Philosophy. Includes a complete discussion of system design, its effectiveness, and the reasons why operating personnel should not make changes without consulting the design engineer.

(b) System Familiarization. Operating personnel not previously involved should be shown the location of all equipment and oriented on using the as-built mechanical drawings, control prints, and the automation point schedule. This is a very important step and can save the operating engineer a lot of time.

(c) System Sequence. An adequate generous amount of time should be set aside to review the system sequence, since systems today are very sophisticated and require a thorough familiarity if they are to be operated effectively.

(d) System Maintenance. All maintenance schedules for each piece of equipment should be reviewed. Warranty obligations of a mechanical contractor and manufacturer warranties do not include preventive maintenance unless specified in the original contract document. This must be performed by the building operator.

(e) Systems Diagnosis. This session covers the symptoms, causes, and corrections. It should give the operating personnel enough knowledge to be able to adequately describe any problems encountered and to take emergency steps to keep afloat until help arrives.

(f) Facility Automation System. It is a part of the commissioning process, but is covered by others.

(g) Continuous Training. Personnel changes are frequent in many operating and maintenance organizations, and items covered in the initial training are often forgotten when not used frequently. Videotape presentations will be effective for retraining.

Some of the items applicable to most training programs are (3):

- Familiarization of personnel with project documentation,
- Description of equipment,
- Description of systems (it is desirable to have representatives of major equipment and systems address their specialties),
- Discussion of design intent,
- Walk-through of a project,
- Start-up procedure,
- Operational procedures,
- Shutdown procedure,
- Emergency procedures,
- Routine maintenance,
- Periodic maintenance,

- Overhaul,
- Factory warranties,
- Spare parts,
- Tools, and
- Hands-on operation.

Unless the owner realizes that all the documents have been turned over to him and that all training is completed, none of us will receive credit for doing our job. The mechanical contractor starts the process with a detailed schedule of the final stages of commissioning and forwards it through the proper channels. The owner ultimately receives this information. At this time, the warranty dates are set and communicated to the owner.

14.7 Benefits of Commissioning (8)

The benefits of proper HVAC system commissioning may be evident in: (a) a system that operates as originally intended, (b) proper documentation of system operation, and (c) operating personnel who are totally knowledgeable of system operation. While this may be a revolutionary concept to some segments of our industry, it is an absolute necessity as systems become more complex or the proper operation of a system is in jeopardy.

14.8 Conclusions

The need for proper commissioning of HVAC systems is apparent when one considers that many properly designed and installed systems fail to live up to their intended functional performance. With proper commissioning and consideration given to IAQ requirements, we can provide:

- Conditions for occupancy of buildings requiring the interaction of all involved in the design and delivery. When proper commissioning is a part of the design and delivery, it is entirely possible that IAQ problems will diminish.

- A checklist for orderly performance of the commissioning process. A well-prepared and coordinated plan will inform all people involved of what to expect. The plan, with necessary documentation, will provide a means to demonstrate that the HVAC system is complete and performs as intended.

- A planned approach by the mechanical contractor for start-up of systems.

- Trained, competent and conscientious personnel to supervise the start-up process.

- Involvement by the design engineer in the commissioning process.

- A complete system start-up.

- Detailed and complete documentation.

- Properly planned owner training.

- Owner participation in training.

- Effective communication between all parties involved.

References

1. ASHRAE Standard 62-1989. "Ventilation for Acceptable Indoor Air Quality." Atlanta, GA, 1989.

2. Sterling, E. M. "Designing Healthy Buildings - The Architect's Role in the Commissioning Process." *ASHRAE Transactions*, Vol. 95, 1989.

3. ASHRAE Guideline 1P: Guideline for Commissioning of HVAC Systems, Atlanta, GA, 1989.

4. Lawson, C. N. "Commissioning: The Construction Phase." *ASHRAE Transactions*, Vol. 95, 1989.

5. Trueman, C.S. "Commissioning: An Owner's Approach for Effective Operations." *ASHRAE Transactions*, Vol. 95, 1989.

6. Stone, D.T. "The HVAC Commissioning Plan." *ASHRAE Transactions*, Vol. 95, 1989.

7. Gill, K.E. "Specifying HVAC Systems Commissioning." *ASHRAE Transactions*, Vol. 95, 1989.

8. Brickman, H. "Commissioning: Why We Need It - What are the Benefits." *ASHRAE Transactions*, Vol. 95, 1989.

15. INVESTIGATION OF INDOOR AIR QUALITY IN BUILDINGS

Milton Meckler, P.E.
President, The Meckler Group
Encino, California

15.1 Introduction

Modern buildings are considered to be generally safe and healthy working environments. However, the introduction of new types of pollutants by new synthetic materials, some energy conservation measures that have minimized the infiltration of outside air, tightly-sealed buildings preventing pollutants from escaping, low quality design, construction, and operation and maintenance have contributed to the buildup of indoor air contaminants. The purpose of this chapter is to explore the prevention and correction options available, and establish procedures to follow in case of indoor air quality (IAQ)-related problems. The procedures outlined in this chapter may not need to be followed completely in all investigations. It is intended to be used as a guide for investigating IAQ problems.

15.2 Prevention and Correction

Although some causes of IAQ problems cannot be foreseen and some cannot be corrected without demolition, prevention involves common sense and the majority of causes can be easily and cost-effectively prevented or corrected. Special attention must be paid to complaints of people; potential sources of indoor air pollution; an adequate supply of pollution-free make-up air; and a balanced heating, ventilating, and air-conditioning (HVAC) system distribution and exhaust. Although preventive as well as corrective measures may, in some cases, increase the construction and occupancy costs, inadequate IAQ measures can be more costly in terms of loss of productive time and removal and replacement of materials to correct the problems. Causes of IAQ problems during the design, construction, close-out, operation and maintenance, and occupancy phases may be eliminated or corrected by the following procedures, where applicable.

15.2.1 Design Phase

(a) Carefully choose building materials, finishes, and equipment.

(b) Design an energy-efficient HVAC system that provides a reasonably pollution-free environment. It should be noted that an HVAC system adequately designed for thermal requirements may not necessarily provide a pollution-free environment.

(c) Allow rooms for access for proper maintenance of HVAC systems.

15.2.2 Construction Phase

(a) Do not substitute materials or modify the design which could result in increased indoor air pollution.

(b) Check to ensure that entire HVAC system is installed in accordance with the original design. Note that it is not uncommon for entire runs to be omitted or dead-ended; all moving parts not to be in operation; fans and motors to be omitted; and supply and return ducts to be omitted, interchanged, or interconnected.

15.2.3 Close-Out Phase

(a) Thoroughly clean the HVAC system, especially for dust and debris.

(b) Exhaust emissions from new materials before gasses permeate other building materials and lengthen the off-gassing period. During the initial off-gassing period, high concentrations of formaldehyde and other indoor air pollutants can cause eye, nose and throat irritation, headache, allergic reactions, fatigue, skin rash, and nausea.

(c) Thoroughly balance the HVAC system. Be sure to take air flow measurements as part of the HVAC system balancing process. Consideration should be given to increasing make-up air considerably for the first few weeks, or even months. Occupants should be advised as to potential causes of stress related to moving, odors due to new construction, and person(s) to contact with in case of complaints.

(d) The following test procedure is recommended as a minimum and should be modified, where required, based on the complexity of each building to obtain best possible test results.

1. Inspect the HVAC and electrical systems to verify that:

    - HVAC system has been finalized in accordance with contract requirements, and

- Electrical systems have been finalized in accordance with contract requirements.

2. Clean building interior and ensure that it is free of standing water before testing.

3. Install movable screens, furniture, and other fixtures prior to testing.

4. Set the HVAC system to operate at a maximum heat mode for at least 24 hours, with full illumination. Time required to heat structural mass may be more than 24 hours.

5. Set the HVAC system to operate at a normal operation mode for 12 to 24 hours after a predetermined temperature has been reached.

6. Set the HVAC system to operate at a maximum heat mode for 12 to 24 hours with full illumination after a predetermined temperature has been reached.

7. Set the HVAC system to operate at a normal mode for a minimum of 24 hours after a predetermined temperature has been reached.

(e) In a building with repetitive floors and separate HVAC systems, it may be economical to measure the IAQ levels on the first floor in a week or two to determine if the process is necessary, or that the process is effective. This will be valuable information in case of litigation.

15.2.4 Occupancy Phase

(a) Heat (or cool) at 6:00 am daily with a 100% air change for the first weeks or longer if necessary. Then heat (or cool) and ventilate normally. Typically, the daily cycle begins just before the arrival of the first employee who finds stagnant air that has not had a 100% air change since the last employee left, and has been absorbing indoor air pollutants for the last 12 hours.

(b) For the first few months, run the HVAC system on the "cool side" during hours of occupancy. This will help keep off-gassing at a minimum during those hours.

(c) Assign one regular employee on each floor to fill out a daily IAQ evaluation form. Two or three times daily, the employee should report his subjective evaluation of temperature, humidity, stuffiness, etc., and any other comments he may have.

(d) Gradually decrease the special heating/ventilation mode in three to six months while monitoring employee comments until the normal heating/ventilation mode is reached.

15.2.5 Operation/Maintenance Phase

(a) Thoroughly train personnel responsible for their portion of operation and maintenance of all building systems.

(b) Avoid contamination of outside air intakes. This type of contamination may be caused by standing water, nests or feathers near outside air intake, and also polluted outside air from adjacent buildings, parking garages, streets, or freeways. Among the other contamination sources to avoid are unfiltered outside air intakes or dirty outside air filters.

(c) To avoid buildup of indoor air pollutants, use adequate ventilation. Decrease in lighting load to conserve energy may result in decreased heating load and air changes.

(d) Avoid inadequate make-up air (volume of outside air introduced) that will cause excessive recirculation of indoor air pollutants. Increased heating and cooling loads will result in decreased make-up air.

(e) Avoid inadequate air distribution (circulation) that will create stagnant air. Furnishings or equipment may also interfere with proper circulation of air.

(f) Keep moving parts of the HVAC system operational at all times.

15.3 Procedures to Follow in Case of Indoor Air Quality Problems

The following general procedures are recommended to be used by the building owners, managers, operators, and others should the need arise to conduct investigations resulting from occupant complaints in their buildings.

15.3.1 Emergencies

(a) In case of an emergency for medical attention or assistance in evacuating a building due to people experiencing severe reactions to some airborne substance, be thoroughly prepared to contact local ambulance, fire and police departments without panic.

15.3.2 Information Required to Investigate Indoor Air Quality Problems in a Building

The following list of questions is designed to assist in gathering information about the occupants, indoor environment, and the ventilation system of a building. In addition to the applicable questions in this list, other questions may be selected by the responsible party where needed.

(a) Inquiries About Occupants:

1. How many people (or what percentage of those asked) now have complaints?

2. Which groups of occupants have the majority of complaints?

3. Were the complaints initiated by a person or a group of people?

4. How consistent are the reported symptoms or complaints?

5. Do occupants complain about the same types of symptoms at home, weekends, holidays, and vacations?

6. Are the symptoms continuous or do they occur at a specific time or with respect to some specific activity in or near the building?

7. When did the problem begin? If the complaints began at a specific time, what was going on then that might have caused or contributed to the problem?

8. Has anyone consulted a physician and if so, what were the diagnosis and treatment?

9. Did occupants experience any problems in their former buildings?

(b) Inquiries About Building Environment:

1. Are there any potential sources of contamination or equipment suspected of causing a problem?

2. Are the problems confined to certain parts of the building?

(c) Inquiries About Building Ventilation System:

1. Does the outdoor air introduced into the building (in cfm/person or cfm/sq. ft. or floor area) satisfy the originally designed rates and ASHRAE recommendations for this type of building?

2. How well is the ventilation system maintained?

15.3.3 Site Inspection

To isolate the problem correctly, further information may be necessary and should be collected by a "walk-through" process looking for possible indoor air contaminants in building materials, finishes, furnishing, equipment, and supplies. In some cases, several site inspections and a more intensive investigation including environmental monitoring may be necessary.

(a) Initial Walk-Through Inspection

1. Prior to Initial Inspection:

   - Review the list of employees, blueprints of the building, blueprints of the HVAC system, modification records of the HVAC system, building operation and maintenance records, etc.

2. Day of Initial Inspection:

   - Bring along a camera, measuring tape, calculator, flashlight, ladder, tool kit, smoke tubes, airflow meter, carbon dioxide monitor, coveralls, thermometer, psychrometer, etc. and be discreet.

   - Assemble everyone involved and explain what will be done. Schedule a meeting later in the day to let them know when you will be able to report on your findings.

- Interview the building occupants carefully. It is best to conduct interviews in private so that people feel free to discuss their concerns openly. Question the occupants about their symptoms, duration of the problem, and what they believe to be the cause.

3. Examination of HVAC Systems

Particularly, examine the following components of the HVAC system in the building:

- Outdoor Air Intakes. Examine outdoor air intakes to determine whether the dampers are shut or blocked. Check for contamination sources and building exhaust near intakes.

- HVAC Equipment. Inspect to determine if the HVAC equipment to condition, treat, and move the air are reasonably clean, dry and running smoothly. Inspect the air filters. Check for fouling and excessing accumulation of dirt and soot.

- Ductwork. Check inside the ductwork to see if it is clean and free of debris.

- Supply Air Diffusers. Thoroughly check all supply air diffusers in the building. Make sure all supply air diffusers are reasonably clean and air is coming out of each one.

- Return Air Grilles. Locate the return air grilles or plenums for each floor or room. Make sure each return air grille draws air from the area.

15.3.4 Bake-Out

Increasing the ventilation rate in new or renovated buildings may not always be sufficient to reduce concentrations of the volatile organic compounds (VOCs) produced by off-gassing of the new building materials and furnishings to acceptable levels. Control of these pollutants may require novel techniques such as conducting a "bake-out." A bake-out is a process of simultaneous or alternative application of heating and ventilation to increase the emissions of VOCs from building materials and remove them from indoor air. In this procedure, the building air temperature is elevated for a certain period of time while continuing to provide ventilation. If the temperature is high enough and of sufficient duration, the residual solvents from materials are driven off, in effect, prematurely by aging the building materials and furnishings.

This is still an area of active research. However, the IAQ Program of the California Department of Health Services has made some observations based on their study of the bake-out of five buildings. It appears that to obtain significant reductions of VOCs, a building must be baked-out for at least three days at a temperature of 90°F to 100°F. Some ventilation must be provided during the bake-out to flush out the pollutants off-gassed; this ventilation rate does not need to be large. Before a bake-out is conducted, an assessment should be made regarding possible material damage because of high temperatures and low relative humidity. Because the procedure will increase costs and delay occupancy of the building, its use may be appropriate only in certain cases, such as when the building's new tenants have heightened concerns about IAQ or when there is reason to suspect that indoor air pollution in a particular building may be much higher than normal.

15.3.5 Further Investigation

An extensive investigation may be necessary in cases where it is impossible to isolate the cause or source of a problem by a visual inspection. In such cases, the following steps are highly recommended:

(a) In the HVAC system, measure the actual air flow rates (in cfm) of fresh and recirculated air in the building, and compare with design airflow rates and recommended airflow rates of ASHRAE Standard 62-1989(1).

(b) Measure the indoor air temperature and relative humidity to see if they are between 23 °C and 25 °C (73 °F and 77 °F) and between 40% and 60%, respectively.

(c) Measure and compare the outside air temperature, air temperature exiting mixing chamber (before heated or cooled), and building return air. There may be inadequate fresh air supply if the mixing chamber air and building return air temperatures are similar, but different than outside air temperature.

(d) Measure indoor air quality (air-sampling) as a last resort. It is expensive and results are not always easily interpreted. Only a few indoor air pollutants are measured inexpensibly. In cases where an outside consultant is needed, select an expert consultant to investigate a specific case. Various environmental consultants may specialize in only one or two of the many disciplines required for a meaningful investigation. Some of the most commonly collected samples are: carbon dioxide, combustion byproducts (nitrogen dioxide or carbon monoxide), particulates, formaldehyde, asbestos, radon, biological contaminates, and VOCs.

In taking air quality measurements, the following common rules are suggested:

- Collect samples outside, near HVAC air intakes, areas where people are experiencing problems and in a control (i.e., non-complaint but otherwise similar) area.

- Take baseline measurements outdoors and indoors early in the day, and repeat them when the building is occupied with the ventilation system running for a long time.

The Occupational Safety and Health Administration (OSHA) has developed sampling and analytical procedures to meet precision and accuracy requirements for airborne contaminants in the range of OSHA permissible exposure limits (PEL) and American Conference of Governmental Industrial Hygienists (ACGIH) threshold limit values (TLV). These procedures are used for sampling eight-hour, time-weighted average (TWA) and short term exposure limit (STEL) of 15 minutes. Sampling procedures chosen for IAQ purposes must be capable of determining concentrations of toxic materials much lower than those normally found in industrial investigations. A few procedures have been validated for these lower level contaminants.

Screening samples may be collected using detector tubes or direct reading instruments. For screening, higher flow rates or longer sampling times may be used for increased sensitivity. Low range detector tubes are available from manufacturers. Based on screening results, validated sampling procedures may be required to further quantify occupant exposures. Much of the information on validated sampling and analytical methods is contained in the OSHA Chemical Information Manual or OSHA Analytical Methods Manual. The following is a short list of direct reading sampling equipment and screening techniques for common indoor air contaminants.

Carbon Dioxide. Low-level detector tubes (0 to 2000 ppm) or portable infrared spectrometers can be used. Carbon dioxide measurements are a useful screening technique in determining whether adequate outside fresh air has been introduced and distributed into the building. The National Institute of Occupational Safety and Health (NIOSH) has made the following recommendations based on the following carbon dioxide concentrations:

- 250 ppm to 350 ppm - normal outdoor ambient concentrations
- Less than 600 ppm - minimum air quality complaints
- 600 ppm to 1000 ppm - less clearly interpreted

- Greater than 1000 ppm - indicates inadequate ventilation. One thousand ppm should be used as an upper limit for IAQ. If carbon dioxide levels exceed 1000 ppm it does not necessarily mean the building is hazardous and should be evacuated. This concentration should be used as a guideline.

Carbon Monoxide. Low-level detector tubes (2 ppm to 200 ppm) or direct reading carbon monoxide monitors can be used.

Formaldehyde. Low-level detector tubes (0.04 ppm to 1 ppm) may be used to evaluate complaints due to off-gassing from insulation, building materials, carpets, drapes, or glues and adhesives.

Nitrogen Oxides and Ozone. They can be screened by using detector tubes. Outdoor samples should also be collected since ambient levels of ozone may reach concentrations that are one to three times the PEL of 0.1 ppm during air inversions. If a more accurate or continuous ozone evaluation is required, a chemiluminescent monitor can be used to measure in the range of 0.01 ppm to 10 ppm. This is a complex instrument that must be calibrated and requires ethylene gas to operate.

Radon. A rapid and easy screening method (Radon Cartridge) for measuring radon concentrations is available. It is used to determine if additional measurements are required.

The Environmental Protection Agency (EPA) makes the following recommendations:

- Less than 4 pCi/L of air - follow-up measurements probably not required

- Greater than 4 pCi/L - follow-up measurements should be performed.

According to the EPA, indoor radon concentrations can be reduced to about 4pCi/L. For levels in the range of 4-20 pCi/L, there is a concern about long-term exposure. For the range of 20-200 pCi/L, an additional 12 months during which precautions are taken to reduce exposure could cause a significant increase in health risk.

Airborne Particulates. A particle counting instrument capable of measuring particulate concentrations as low as 2000 particles/cc of air can be used.

Airborne Microorganisms. The ACGIH recommends a preassessment of microbial contamination prior to air-sampling. Several precautions must be taken in preparing culture media for sampling, specialized handling and shipping procedures, and analyzing by a laboratory familiar with handling and processing of biological samples. The following criteria may be used as a base for investigation:

- 1000 viable colony forming units in a cubic meter of air
- 1,000,000 fungi per gram of dust or material
- 10,000 bacteria or fungi per milliliter of stagnant water or slime.

Concentrations in excess of the above do not necessarily imply that the conditions are unsafe or hazardous. The types and concentrations of the airborne microorganisms will determine the hazard to the occupants.

Miscellaneous Contaminants. A portable infrared spectrometer can be used to evaluate a wide range of potential indoor air contaminants including acetic acid, ammonia, carbon dioxide, carbon monoxide, nitric oxide, nitrogen dioxide, sulfur dioxide, and VOCs. Care must be taken in analyzing and interpreting the results since the instrument is not always specific for one compound.

15.3.6 Final Report and Follow-Up

A final report may be needed. This should include the observations, summary of findings, the results of measurements made or samples collected, and their significance to the problem. Make certain that all occupants are kept informed of your progress. Reevaluation of a building after repairs or modifications have been made may be advisable to check which control measures work best for eliminating IAQ problems.

References

1. ASHRAE Standard 62-1989. "Ventilation for Acceptable Indoor Air Quality." Atlanta, GA, 1989.

INDEX

B

C

D

E

F

S

T

U

V

W

Z